Paläoklimaforschung

Akademie der Wissenschaften und der Literatur

Paläoklimaforschung
Palaeoclimate Research Volume 9

Special Issue: ESF Project
"European Palaeoclimate and Man" 4

Editor: Burkhard Frenzel

Associate Editor: Birgit Gläser

1993

European Science Foundation
Strasbourg

Akademie der Wissenschaften
und der Literatur · Mainz

Oscillations of the Alpine and Polar Tree Limits in the Holocene

Edited by Burkhard Frenzel

Co-edited by Matti Eronen, Karl-Dag Vorren & Birgit Gläser

81 figures and 8 tables

Gustav Fischer Verlag · Stuttgart · Jena · New York · 1993

Der vorliegende Sonderband wurde mit Mitteln der European Science Foundation (Straßburg) und der Akademie der Wissenschaften und der Literatur (Mainz) gefördert. Die Verantwortung für den Inhalt dieser Veröffentlichung liegt bei den Autoren.

Anschriften der Herausgeber:

Prof. Dr. Dr. h.c. Burkhard Frenzel, Institut für Botanik 210 der Universität Hohenheim, Garbenstraße 30, D-7000 Stuttgart 70, F.R.G.

Dr. Matti Eronen, Department of Geology, University of Oulu, Linnanmaa, 90570 Oulu, Finland

Dr. Birgit Gläser, Institut für Botanik 210 der Universität Hohenheim, Garbenstraße 30, D-7000 Stuttgart 70, F.R.G.

Prof. Dr. Karl-Dag Vorren, University of Tromsø, N-9000 Tromsø, Norway

Sprachliche Beratung: B. A. Carolyn Ostridge
Bibliographische Bearbeitung: Dipl. Agr.-Biol. Mirjam Weiß, Hohenheim/ESF
Graphische Bearbeitung: Erika Rücker, Hohenheim
Technische Redaktion: Dr. Birgit Gläser, Hohenheim/ESF

Die Deutsche Bibliothek – CIP Einheitsaufnahme

Oscillations of the alpine and polar tree limits in the holocene
: 8 tables / ed. by Burkhard Frenzel. Co-ed. by Matti Eronen
... - Stuttgart ; Jena ; New York : G. Fischer, 1993
(Palaeoclimate research ; Vol. 9 ; ESF project "European palaeoclimate and man" ; Special issue 4)
ISBN 3-437-30735-5 (Stuttgart ...)
ISBN 1-56081-374-1 (New York ...)
NE: Frenzel, Burkhard [Hrsg]: Paläoklimaforschung / ESF project "European palaeoclimate and man"

Gesamtherstellung: Rheinhessische Druckwerkstätte, Alzey. Printed in Germany.

ISBN 3-437-30735-5
US-ISBN 1-56081-374-1
ISSN 0930-4673

CONTENTS

Addresses of the authors

Prof. Dr. Brigitta Ammann, Systematisch-Geobotanisches Institut, Universität Bern, Altenbergrain 21, CH-3013 Bern, Switzerland

PD Dr. Conradin A. Burga, Geographisches Institut der Universität Zürich-Irchel, Winterthurerstraße 190, CH-8057 Zürich, Switzerland

Dr. Matti Eronen, Department of Geology, University of Oulu, Linnanmaa, SF-90570 Oulu, Finland

Dr. Sheila Hicks, Department of Geology, University of Oulu, Linnanmaa, SF-90570 Oulu, Finland

Prof. Dr. Friedrich-Karl Holtmeier, Institut für Geographie, Abteilung Landschaftsökologie, Westfälische Wilhelms-Universität Münster, Robert-Koch-Straße 26, D-4400 Münster, F.R.G.

Dr. Joachim Hüppe, Universität Hannover, Institut für Geobotanik, Nienburger Str. 17, D-3000 Hannover 1

Dr. Pertti Huttunen, University of Joensuu, Karelian Institute, P.O. Box 111, SF-80101 Joensuu, Finland

Dr. Hannu Hyvärinen, Department of Geology, University of Helsinki, Snellmaninkatu 5, SF-00100 Helsinki, Finland

Dr. C. Jensen, University of Tromsø, N-9000 Tromsø, Norway

Prof. Wibjörn Karlén, Department of Physical Geography, University of Stockholm, P. O. Box 6801, S-113 86 Stockholm, Sweden

Dr. Leif Kullman, Department of Physical Geography, University of Umeå, S-901 87 Umeå, Sweden

Dr. Mons Kvamme, Botanical Institute, University of Bergen, Allégaten 41, N-5007 Bergen, Norway

Dr. John A. Matthews, Department of Geology, University of Wales, College of Cardiff (UWCC), P.O. Box 914, Cardiff CF1 3YE, Wales, U.K.

Dr. R. Mook, University of Tromsø, N-9000 Tromsø, Norway

Dr. B. Mørkved, University of Tromsø, N-9000 Tromsø, Norway

Prof. Dr. Richard Pott, Institut für Geobotanik, Nienburger Straße 17, D-3000 Hannover 1, F.R.G.

Dr. Kamil Rybníček, Institute of Systematic and Ecological Biology, Czechoslovak Academy of Sciences, Květná 8, CS-603 65 Brno, Czechoslovakia

Dr. Eliška Rybníčková, Institute of Systematic and Ecological Biology, Czechoslovak Academy of Sciences, Květná 8, CS-603 65 Brno, Czechoslovakia

Prof. Dr. Fritz Hans Schweingruber, Swiss Federal Institute for Forest, Snow and Landscape Research (WSL), CH-8903 Birmensdorf, Switzerland

Dr. Stepan G. Shiyatov, Laboratory of Dendrochronology, Institute of Plant and Animal Ecology, Ural Division of the Russian Academy of Sciences, 8 Marta Street, 202 Ekaterinburg, 620219 GSP-511, Russian Federation

Prof. Dr. Bjartmar Sveinbjörnsson, Institute of Biology, University of Iceland, Grensásvegur 12, 108 Reykjavik, Iceland

Dr. T. Thun, University of Trondheim, N-7050 Dragvoll, Norway

Prof. Dr. Walter Tranquillini, Institut für Botanik, Universität Innsbruck, Sternwartestraße 15, A-6020 Innsbruck, Austria

Prof. Dr. Karl-Dag Vorren, University of Tromsø, N-9000 Tromsø, Norway

Dr. Lucia Wick, Systematisch-Geobotanisches Institut, Universität Bern, Altenbergrain 21, CH-3013 Bern, Switzerland

Preface

Matti Eronen, Burkhard Frenzel & Karl-Dag Vorren

In early 1989 the European Science Foundation launched the research project "European palaeoclimate and man since the last glaciation". This international and multidisciplinary project aims at detecting any possible influence pre- and early historic man might have had on European climate by forest clearing, agriculture, animal husbandry, and early metallurgy, long before the onset of industrialization.

In case that such an early human impact on regional climates of Europe can really be evidenced, these anthropogenic influences have to be dated and quantified as accurately as possible. On the background of spontaneously changing climates any potential human influence on climate in Holocene times is confined to approximately the last 7000 years of the recent past.

Thus, when trying to differentiate between spontaneously changing climatic conditions and anthropogenically-triggered climatic changes, we are facing a Gordian knot which we are challenged to cut. This seems to be possible only when the history of European Holocene climate is first analysed in those regions which were too remote to be affected by direct or indirect influences of man's activities during the time period in question. These regions are most of all the Greenland inland ice, the North Atlantic and Polar Seas, the tundra zone and the alpine belts of the European high mountains. In a second step the changing Holocene climate has to be studied qualitatively and quantitatively in regions which were strongly influenced by pre- and early historic man during various phases of the past.

It is hoped that in comparing these two data sets, the presumably natural one and the anthropogenically influenced one which was derived from investigations in more or less densely inhabited regions, a possible influence of man on the European Holocene climate can be traced and eventually evaluated.

Oscillations of the polar and alpine tree limits and changes in the width and density of the annual rings of long-living trees thriving in these most important plantgeographical boundary regions contain a wealth of palaeoclimatological information, provided that these parameters can be studied in detail, that their changes can be reliably dated, and that the controlling plantphysiological and plantecological factors of these tree lines are comprehensively understood. Since it is evidently of scientific importance to focus on these problems in detail, specialists from various European countries were invited by the European

Science Foundation to a workshop on "Oscillations of alpine and polar tree lines in the Holocene" (Innsbruck, Austria, 19-21 October 1990) to discuss the following topics, and the interrelations between them:

(1) Recent tree lines and their climatic implications;
(2) The connection between present and past tree line ecology, how to record past tree lines and their fluctuations, and how to use the data in palaeoclimatic reconstructions;
(3) Palaeoecological approaches and results in different regions from:
  (a) Tree-ring and megafossil investigations;
  (b) Stratigraphical investigations.

After the paper sessions the symposium discussions concentrated on:

(1) The value of tree line oscillation data in palaeoclimatic reconstructions;
(2) Critical examination of tree-rings as climate proxy data;
(3) How to correlate stratigraphical and glacier data with the tree-ring curves;
(4) How to improve the temporal resolution of the stratigraphical data.

Three working groups were formed during the conference to pursue the following questions:

(A) What are the interrelations between the position of tree lines and plantecology / meteorology;
(B) What are the links between the structure of tree-rings, climate, and the structure and position of the timberline;
(C) How can "palaeo-treelines" be used in palaeoclimatology:
  (a) Identification of former tree lines by stratigraphical methods;
  (b) Interpretation of palaeoclimate based on stratigraphical data.

The groups generally recommended a concentration on a more basic level as well as on interdisciplinary research. From an ecological/meteorological point of view, there is a need for long-term monitoring experiments which should further investigate the ecophysiology (especially of the root zone) on population level, and on ecosystem level (the microclimates in these ecotones). Investigation in Eastern and the westernmost Europe should be encouraged.

Dendrochronological studies usually have to cope with problems in interpreting thickness variations of tree-rings. Polar and alpine tree line zones have proved to be very suitable areas for tree-ring studies, because there the growth of trees is mainly determined by summer temperatures. As for multidisciplinary efforts, overlapping palaeoclimatic proxy data from tree-rings and sediments are still lacking. In this context, methods aiming at cross-datings with laminated sediments should be developed, and cooperation between dendrochronologists and glaciologists should be initiated. In addition, stable isotope measurements from tree-rings will give valuable palaeoclimatic information which can be used for correlations with other data.

Another point of interest is the establishment of large-scale dendrochronological networks: dense networks of around 300-year-long tree-ring series would enable a considerable improvement in the accuracy of summer temperature reconstructions. Within such a network longer tree-ring series (key series / type series) should also be developed.

At present, only one European laboratory is available for maximum density measurements of tree-rings works on a routine basis. Thus a great need towards the establishment of regional laboratories is felt.

In the field of stratigraphic palaeoclimatic proxy data good type sections of annually laminated sediments are in demand. For further progress interdisciplinary approaches in selected areas, combined with the employment of new methods, should be encouraged. Regional workshops, field and laboratory meetings are recommended.

The workshop served well as a multidisciplinary approach to the problems connected with palaeoclimatic studies in the alpine and polar tree line areas. It showed that there are still many problems in the climatic interpretation of data but at the same time these areas offer good prospects for ongoing and future research with new techniques.

Forest limit investigations in northernmost Finland

Matti Eronen, Sheila Hicks, Pertti Huttunen & Hannu Hyvärinen

General introduction

The area covered by the following three investigations consists of the northern part of Finnish Lapland and the adjacent parts of Norway as delimited in Fig. 1. Within this area, the majority of which belongs to the northern boreal zone (AHTI et al., 1968), the conifers pine (*Pinus sylvestris*) and spruce (*Picea abies*) have their northernmost limits while the deciduous mountain birch (*Betula pubescens ssp. tortuosa*), experiences an altitudinal limit. There is, therefore, a distinct vegetation zonation from the barren treeless mountain tops, through open birch woodland, to pine dominated forests and, eventually, pine forests with a strong admixture of spruce (Fig. 2; SUOMEN KARTASTO, 1988).

However, further north, in Norway, the latitudinal and oceanward limit of the mountain birch is also crossed so that along the outer shores of the fjords barren treeless areas are found at sea level. Locally, however, the situation is reversed and, on descending from the mountain birch woodland zone to the Norwegian coast, pines reappear at several sheltered fjord heads.

In Finland the vegetation of the barrens above the tree line is oroarctic in nature and consists primarily of dwarf shrub heath dominated by *Empertrum nigrum* or *Vaccinium myrtillus*. Dwarf shrubs are absent, however, from areas where strong winds prevent the accumulation of snow in winter and from snow patch areas, where a meadow vegetation exists. The zone of mountain birch woodland is more oceanic in character. The birches themselves are low-growing and frequently have many trunks from the same stock. Juniper is often commonly present as an understorey. Within this birch area, however, there are quite extensive islands of pine. Every now and again considerable areas of the birch forest are destroyed by the caterpillar of the moth *Epirrita autumnata*. The pine dominated forests occupy the most continental part of the country and this, together with the prevalence of sandy soils in this area, means that the forests are often very dry and dominated by lichens. Of interest is the fact that, only here, does a species of pine grow further north than spruce. Everywhere else in the polar zone the situation is reversed, with the exception of the dwarf pine (*Pinus pumila*) in Northeast Asia.

The location of the ecotones between these zones is determined primarily by climate. That climatic conditions have generally become less favourable in the more recent past is seen by the fact that the vast majority of pines present in the birch woodland zone beyond the pine forest limit are large mature individuals, often of great age, and young regenerating trees are common only in the larger pine islands. It is the movement of these forest limits in the past, particularly that of pine, which form the focus of the investigations by ERONEN & HUTTUNEN, HICKS, and HYVÄRINEN which follow.

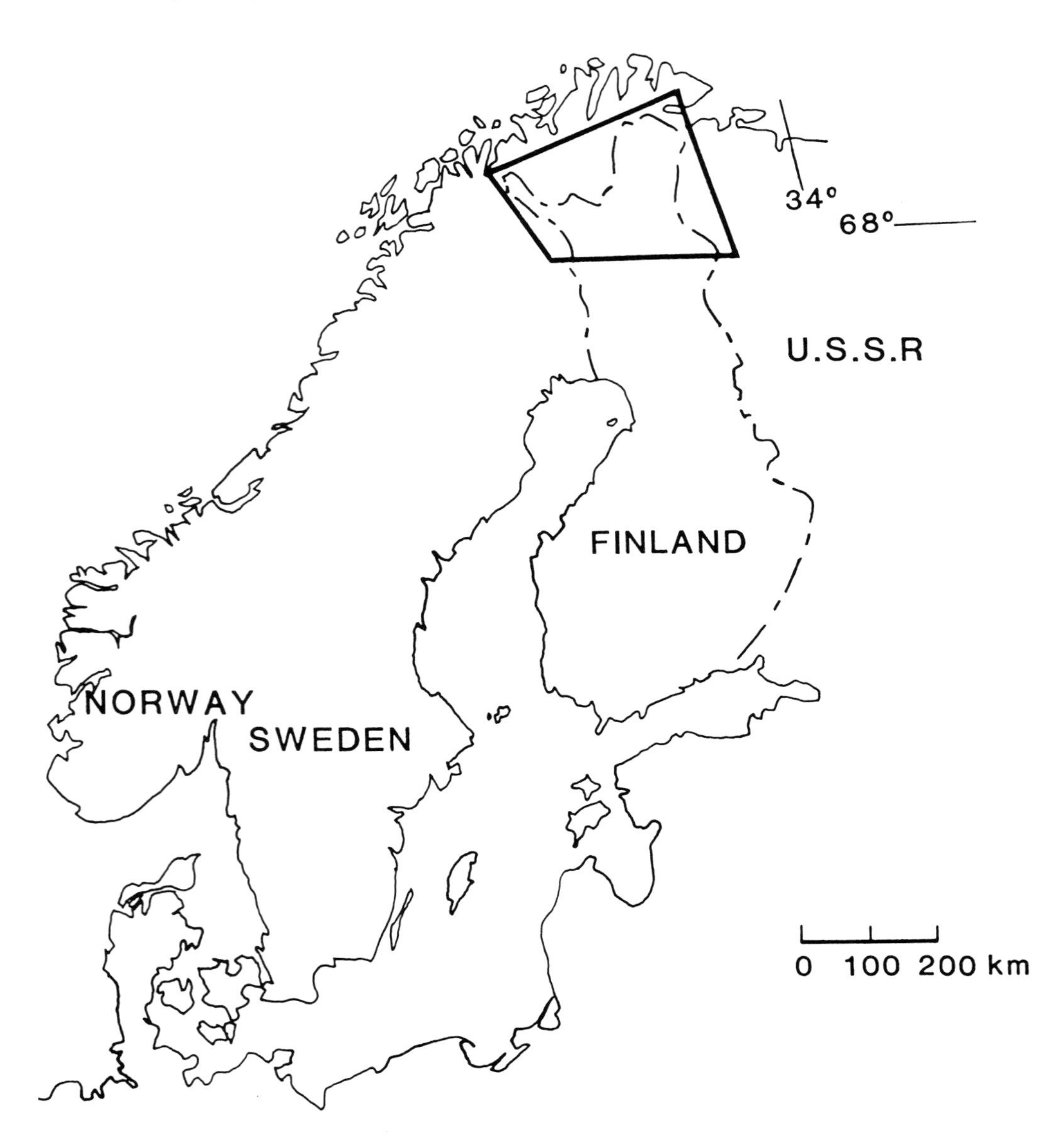

Fig. 1 Fennoscandia showing location of the study area

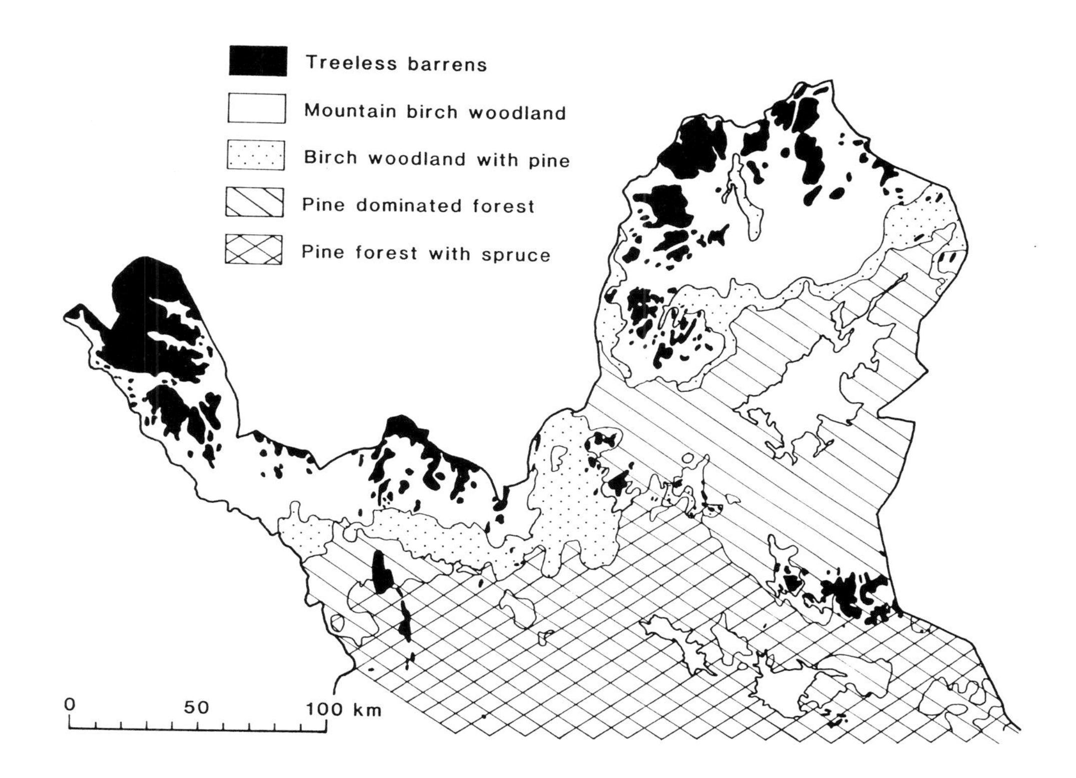

Fig. 2 Major vegetation zones of Finnish Lapland (SUOMEN KARTASTO, 1988)

References

AHTI, T.; HÄMET-AHTI, L & JALAS, J. (1968): Vegetation zones and their sections in Northwestern Europe. Ann. Bot. Fenn. 5, 169-211

SUOMEN KARTASTO (1988): Folio 141-143 Biogeography, Nature conservation

Addresses of the authors:

Dr. M. Eronen, Department of Geology, University of Oulu, Linnanmaa, 90570 Oulu, Finland
Dr. S. Hicks, Department of Geology, University of Oulu, Linnanmaa, 90570 Oulu, Finland
Dr. P. Huttunen, University of Joensuu, Karelian Institute, P.O. Box 111, SF-80101 Joensuu, Finland
Dr. H. Hyvärinen, Department of Geology, Division of Geology and Palaeontology, University of Helsinki, Snellmaninkatu 5, SF-00170 Helsinki, Finland

The use of recent pollen rain records in investigating natural and anthropogenic changes in the polar tree limit in Northern Fennoscandia

Sheila Hicks

Summary

Changes in the forest limit in Finnish Lapland during the Holocene have been primarily in response to changes in climate. However, this area has been continuously occupied by nomadic groups which may also have influenced the nature of the local forest communities. The project aims, by means of pollen analysis, to distinguish between these two types of influence and to ascertain the timing, duration and areal extent of each. Modern pollen rain values will be used to delimit both pollen trends which are related to climate and those pollen features which are diagnostic of different vegetation communities, particularly anthropogenically induced ones. These will then form the basis for a detailed interpretation of the Holocene as recorded in peat deposits within the close vicinity of sites which are known to have been occupied over long periods of time.

Zusammenfassung

Die Veränderungen, die sich während des Holozäns an der Waldgrenze Finnisch-Lapplands abgespielt haben, wurden hauptsächlich durch klimatische Faktoren verursacht. Mit dem Einsetzen einer permanenten Besiedlung durch nomadische Bevölkerungsgruppen können jedoch anthropogene Einflüsse auf die Eigenschaften der lokalen Waldgesellschaften nicht mehr ausgeschlossen werden. Die vorliegende Untersuchung zielt darauf ab, die klimatischen den anthropogenen Einflußgrößen gegenüberzustellen, sie zeitlich festzulegen sowie ihre Dauer und Raumwirksamkeit abzuschätzen. Untersuchungen des heutigen Pollenniederschlags werden genutzt, um Rückschlüsse auf klimatische bzw. anthropogene Faktoren ziehen zu können (Auswertung klimatisch verursachter Pollentrends bzw. anthropogen beeinflußter Pflanzengesellschaften). Auf dieser Basis wird der Versuch einer detaillierten Interpretation des Holozäns unternommen. Als Datengrundlage dienen holozäne Torfablagerungen, die aus unmittelbarer Nachbarschaft nachweislich langbesiedelter Gebiete stammen.

1. Introduction

The main aim of this project, which is only in its initial stage, is to separate out those changes in the forest limits within the area of Lapland described in the preceding introductory section (pp 1-3) which are the result of climatic changes from those changes possibly caused by man. The main tool is pollen analysis, both of Holocene sediments and modern vegetation units. One can hypothesize on the differences in the nature of climatically and anthropogenically induced changes at the polar forest limit and if there is any basis on which the two may be distinguished from each other. The following are suggested:

1.1 Climatically induced changes

(1) regional in effect - revealed in changes in the forest vegetation;
(2) usually a long term trend rather than a sudden event;
(3) slow, if point 2 holds good;
(4) synchronous over a wide area (see point 1).

1.2 Anthropogenically induced changes

(1) may be highly local - seen in changes in the field and ground level vegetation;
(2) can be of very short duration (one season) or a particularly long term event (consistent seasonal occupation for thousands of years);
(3) rapid or slow. Destructive changes can be achieved in a very short time using, for example, axe or, either intentionally or unintentionally, fire. On the other hand small scale but persistent interference can cause slow changes;
(4) timing and duration likely to vary from place to place.

2. Methods and preliminary results

The changes outlined above can be followed by means of pollen analysis. The approach in this project is to monitor present-day situations and use the analogues or "pollen pictures" thus obtained as the basis for interpreting pollen diagrams covering periods in the past. This is feasible as far as comparable situations exist at the present day. It also potentially pinpoints any situations in the past which have no modern counterpart. Although the modern pollen pictures can be built up on an annual basis, the precision with which events from the past have been recorded will, of course, depend on the accumulation rate of the sediments being analysed. If this is slow then short-term changes will go unnoticed.

2.1 Long-term, large scale, synchronous forest changes

Pollen traps modified from the type originally developed by TAUBER (TAUBER, 1965, 1974; HICKS & HYVÄRINEN, 1986) have been used by the author to monitor annual pollen rain within the major forest belts of Northern Finland since 1974. The location of those traps referred to in this paper is shown in Fig. 1. The results of a selection of the traps for the years 1983-84 and 1985-86, expressed in terms of percentages of the total pollen spectrum, are illustrated in Fig. 2 where they are grouped according to the vegetation zones from north to south. There are distinct variations from year to year (see next section) but, at the same time, a clear pollen picture of each vegetation zone emerges.

Open summit

Highest percentage of tundra plants. Shrubs (*Salix* and *Juniperus*), herbs and dwarf shrubs well represented. BUT tree pollen dominates the spectra, *Pinus* in 1983-84 and *Betula* in 1985-86.

Birch woodland

Betula >40% P (higher values than anywhere else except J12). Dwarf shrubs, grasses and sedges common but locally variable.

Pine forest

Pinus >65% P (highest values of any trap). Dwarf shrubs and herbs minimal.

Spruce dominated

Picea values higher than elsewhere but still only 5-10% P. *Pinus* values almost as high as in the pine forest. Dwarf shrubs and herbs more important than in the pine forest but less so than in the birch woodland.

When pollen influx cm^{-2} $year^{-1}$ is considered rather than percentage presence (Fig. 4) it is evident that the highest annual pollen influx is recorded in the pine forest and the lowest on the bare mountain tops. In fact it is the latter feature which accounts for the high arboreal pollen percentages in this open summit situation. The actual amount of *Pinus* and *Betula* pollen is very low and quite insignificant in comparison with the amounts recorded in the forest zones and yet, since this long distance transported pollen is virtually all that is present on the mountain tops, the local plants producing very little pollen, its percentage presence is high. It is obvious, therefore, that the annual pollen influx of each vegetation zone is also characteristic and must be included as part of the pollen picture. In fact it may be a

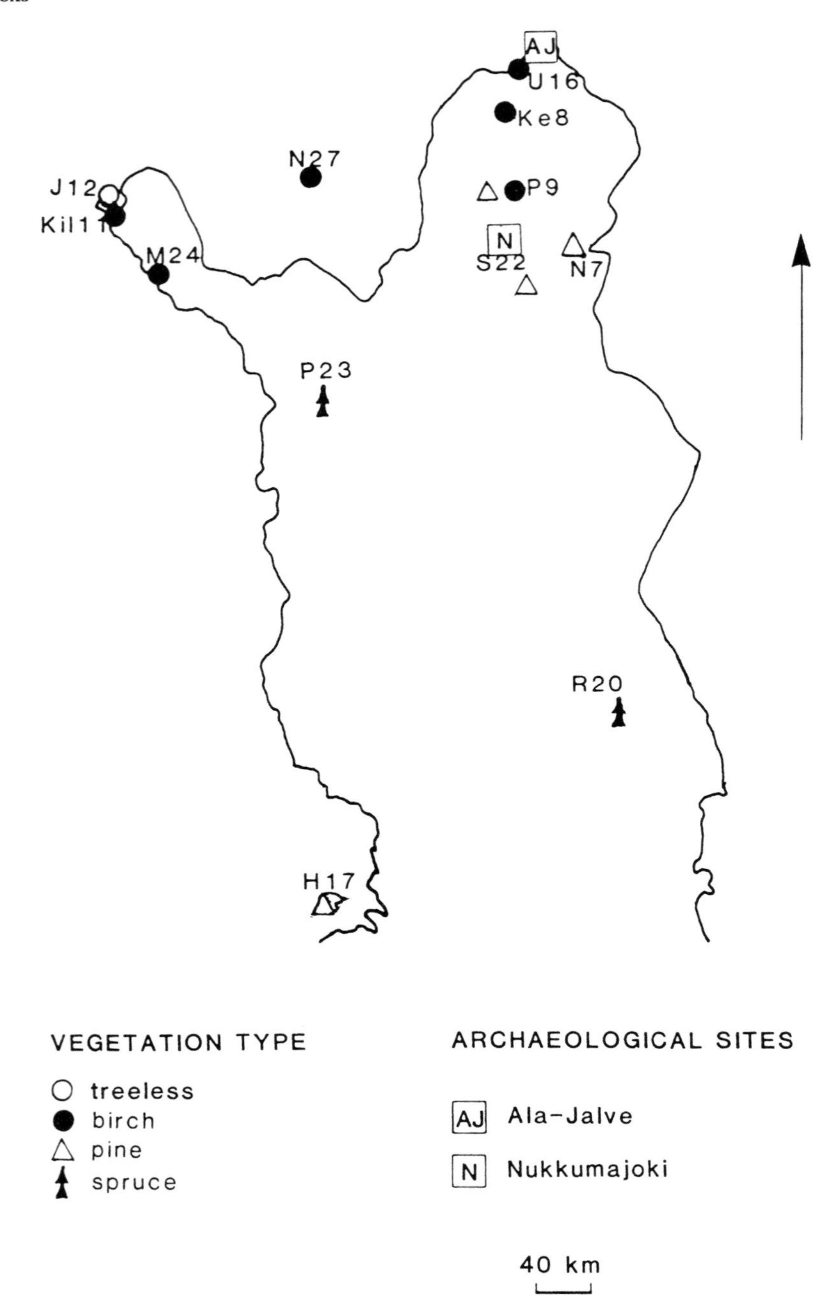

Fig. 1 Northern Finland showing location of pollen traps and the regional vegetation type each represents, together with the two archaeological sites mentioned in the text

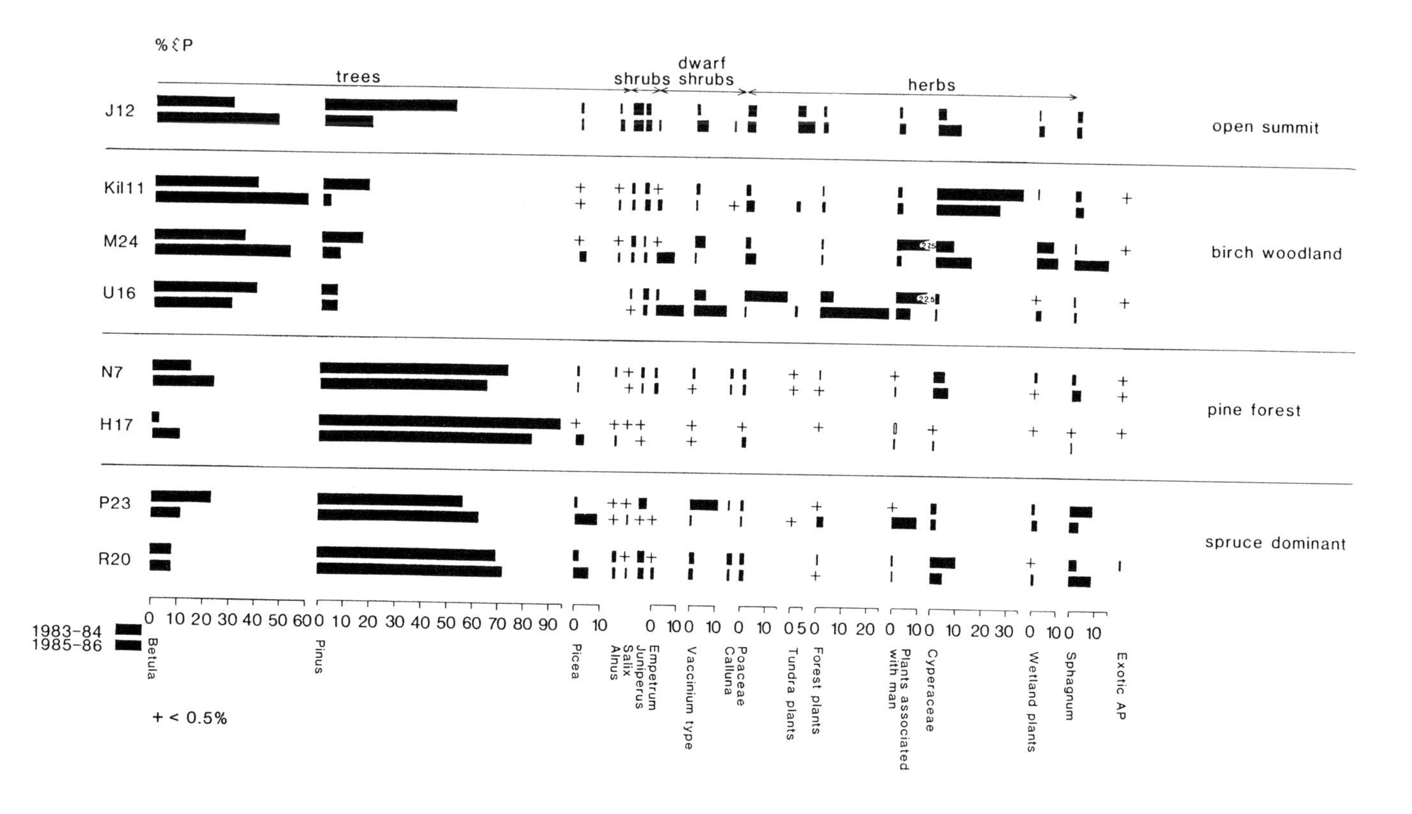

Fig. 2 Pollen trap results illustrative of the different vegetation zones of Northern Finland for the years 1983-84 and 1985-86 expressed as percentages of total pollen

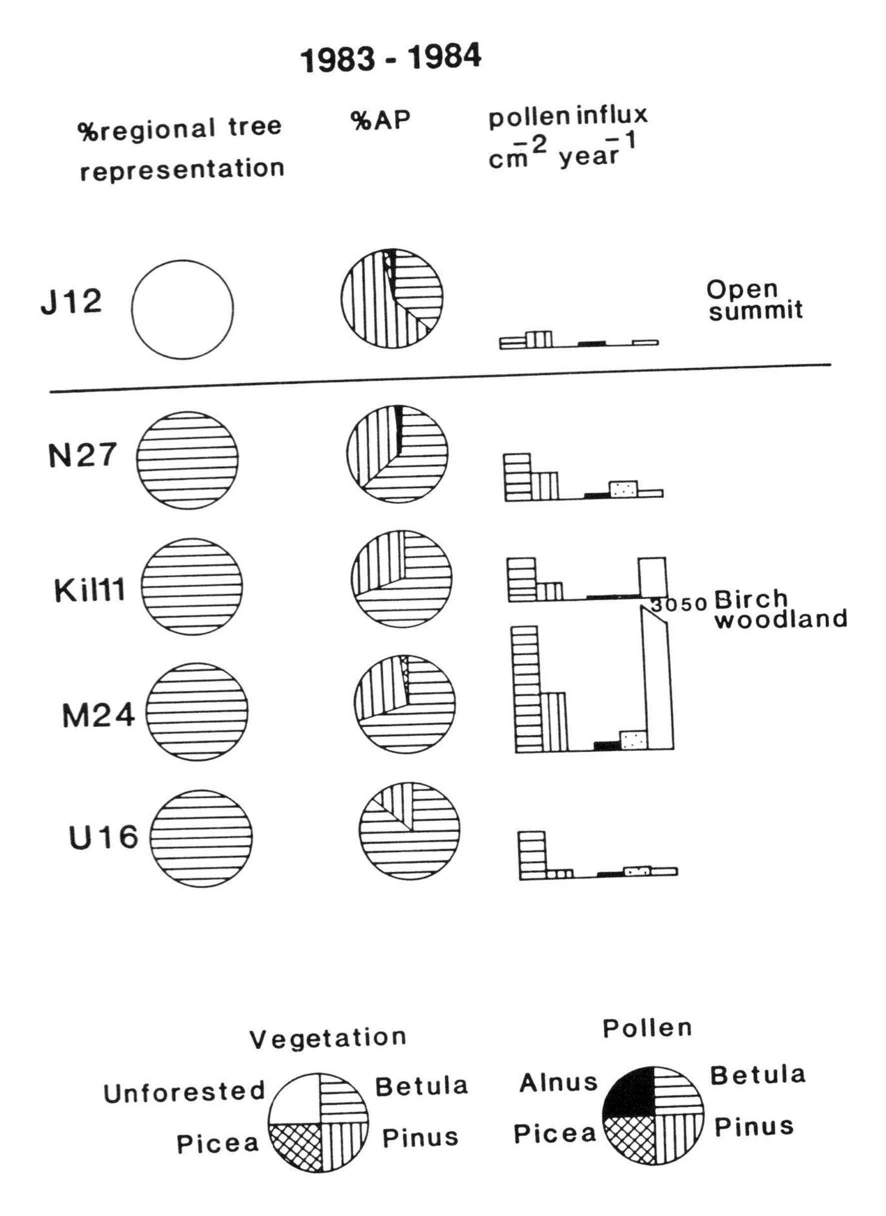

Fig. 3 Relationship between regional vegetation and pollen deposition for the year 1983-84 for the Lapland pollen traps

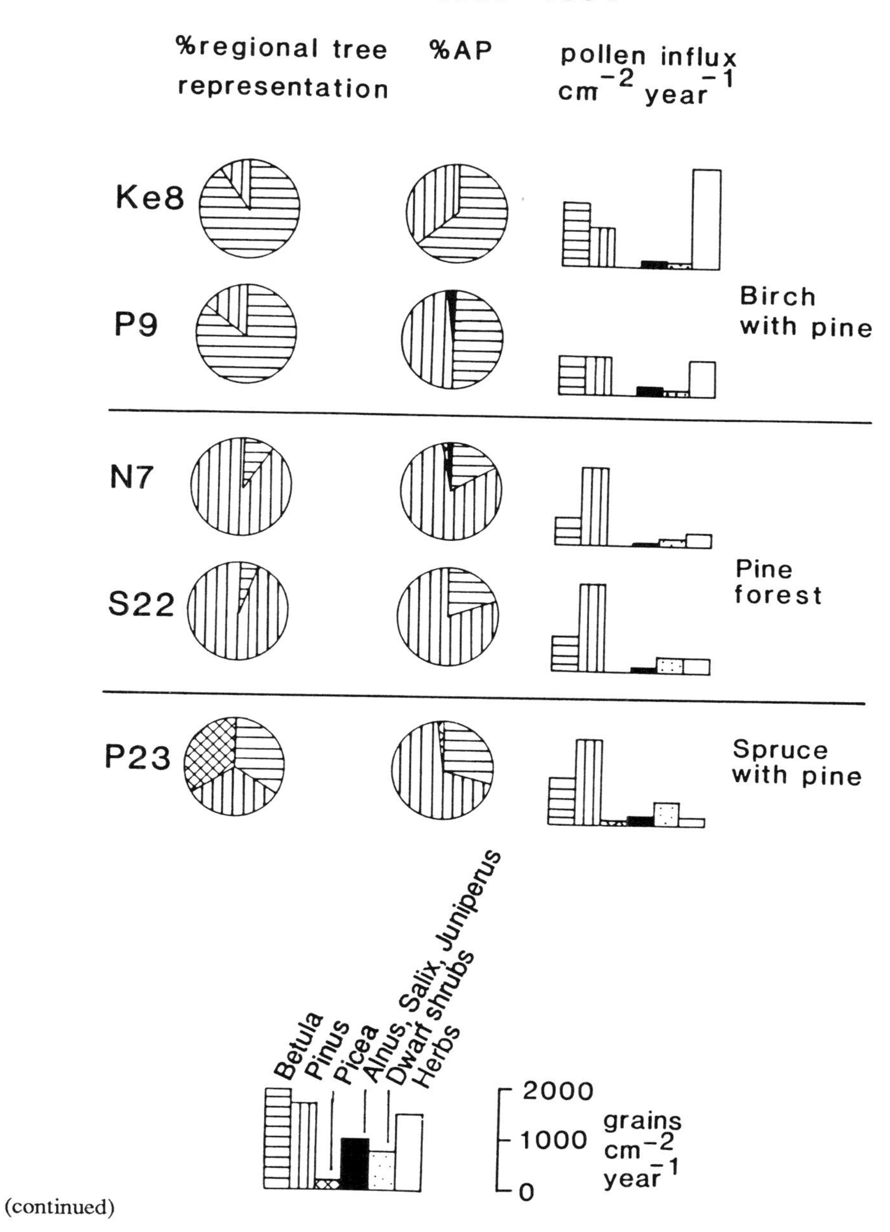
1983 - 1984
%regional tree representation
%AP
pollen influx cm^{-2} year^{-1}
Ke8
P9
Birch with pine
N7
S22
Pine forest
P23
Spruce with pine
Betula
Pinus
Picea
Alnus, Salix, Juniperus
Dwarf shrubs
Herbs
2000
1000
0
grains cm^{-2} year^{-1}

(continued)

crucial factor in pin-pointing movements in the forest limits, as has been amply demonstrated by HYVÄRINEN (1975, 1976). Both the percentage and influx values of the arboreal pollen types can be visually compared with the percentage presence of these same trees in the regional vegetation surrounding the pollen traps by means of Fig. 3. It is clear that there is no 1:1 relationship. *Pinus* pollen is present in the birch woodland even when the tree is absent and *Betula* pollen is present in the pine forest at values far in excess of the tree presence and, as emphasized above, both are present on the unforested summits. Long distance transport is largely responsible although differences in pollen production are also involved. As a general rule, the more open the local vegetation the greater the proportion of pollen which comes from outside (inter al. TAUBER, 1977; PRENTICE, 1985). The fact that there is no 1:1 relationship between pollen presence and tree presence emphasizes the need to monitor the present-day situation and delimit pollen pictures or analogues for the different forest belts so that the pollen assemblages from the past can be interpreted more precisely.

2.2 Short-term changes in pollen production related to climate

Although each vegetation zone presents a clear pollen picture there is, nevertheless, considerable variation in values from year to year, both in percentage and influx terms (Fig. 3 and 4). This is particularly noticeable in the case of *Picea* for which good flowering years occur only rarely in these northern latitudes (KOSKI & TALLQVIST, 1978; HICKS, 1985). This variation is most likely related to climatic and one aim of the present project is to compare the pollen values of the Lapland traps with climatic data over the same period i.e. 1974 to present. In this connection the detailed studies on the behaviour of *Betula* and *Pinus* at the forest limit undertaken at the subarctic research station at Kevo provide an invaluable body of material for supplementing the pollen trap data at this one site. It is hoped that any correlation between annual pollen influx and climate detected in this way may be used to interpret past pollen changes in terms of these same climatic variables. In this respect sediment profiles for which the time scale can be deduced with a high level of precision will be needed. Although laminated sediments would give a yearly record, peats will have to be given priority because the pollen trap values, having been collected for ground surface situations, are only strictly comparable with an equivalent sedimentary environment. Sedimentation on a lake floor involves quite different mechanisms (DAVIS, 1973; DAVIS & BRUBAKER, 1973).

2.3 Small-scale, metachronous vegetation changes caused by man

Modern analogues or pollen pictures can also be obtained for vegetation changes arising as a result of human activity. The strategy here is to locate sites in which the vegetation has been altered as a result of a known activity and, from within them, take a series of moss samples. The mosses act as small pollen traps with the difference that only pollen percent-

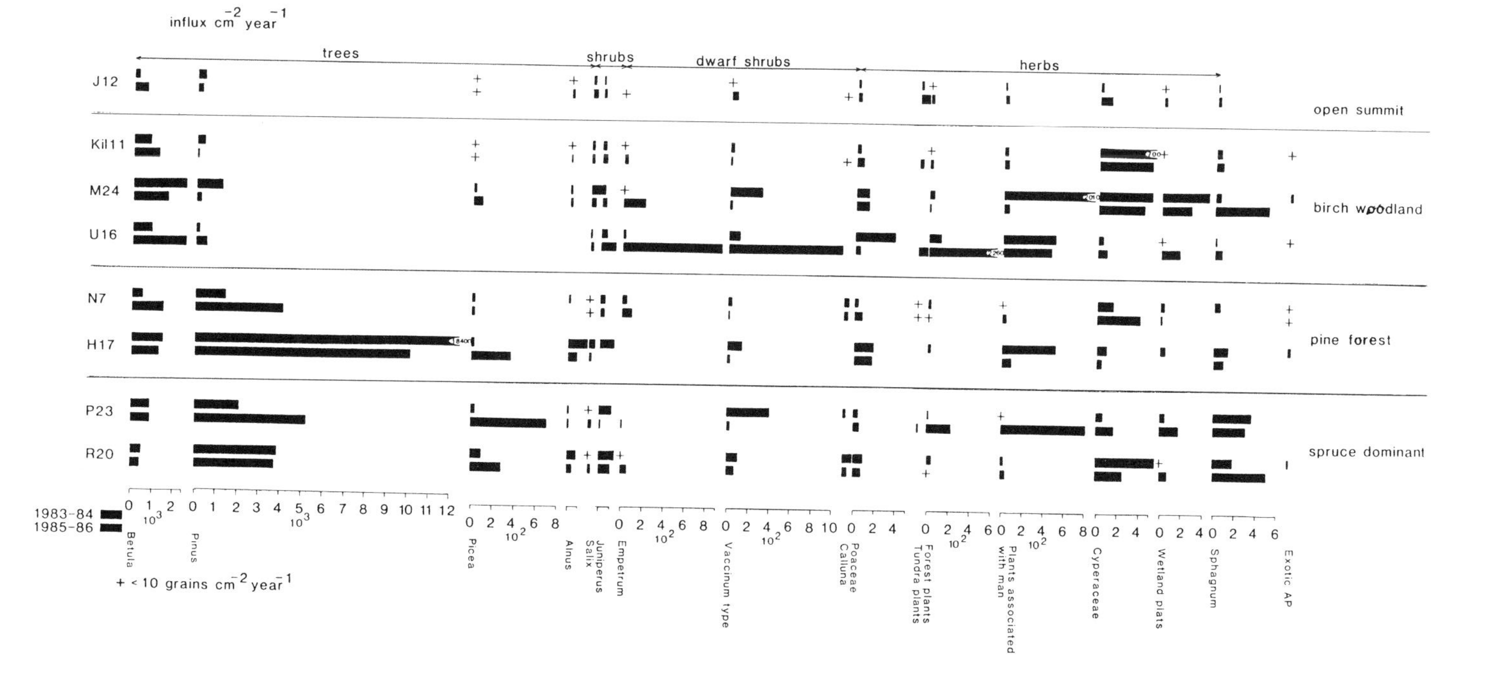

Fig. 4 Pollen trap results illustrative of the different vegetation zones of northern Finland for the years 1983-84 and 1985-86 expressed as pollen influx

ages can be calculated with accuracy from such material not pollen influx ones. One suitable situation is provided by the Lapp winter villages a series of which have been investigated in detail by CARPELAN (CARPELAN et al., 1990) and for which the surrounding vegetation has been analysed by SUOMINEN (1975). The characteristic plants of these villages at Nukkumajoki, just south of Inari (Fig. 1), which were intensively occupied in the fifteenth, early seventeenth and late seventeenth centuries are listed in Table 1, together with an indication of the precision with which each would be identified in terms of pollen. This pollen assemblage is sufficiently distinctive that it should be clearly recognizable in the pollen record. With this in mind the intention is to collect series of moss samples from Lapp winter villages in both the pine and birch zones and build up a pollen picture for each. These can then be employed in analysing peat profiles from the immediate vicinity of known archaeological sites within the same regions.

Table 1 Plants at present indicative of medieval Lapp winter villages at Nukkumajoki, Inari (SUOMINEN, 1975), together with an assessment of their level of identification in pollen terms

	Species	Precision of pollen identification	
consistently present and abundant/common	*Betula pubescens*	Species	
	Juniperus communis	Species	
	Deschampsia flexuosa	Family	Poaceae
	Lycopodium annotinum	Species	
consistently present but less common	*Linnea borealis*	Species	
	Trientalis europaea	Species	
	Calamagrostis lapponica	Family	Poaceae
	Festuca ovina	Family	Poaceae
	Solidago virgaurea	Sub-family	Compositae Solidago type
sporadically present	*Epilobium angustifolium*	Genus	
	Poa rigens	Family	Poaceae
	Diphasiastrum complanatum	Genus	
	Melampyrum pratense	Genus	
	Hierochloe hirta (odorata)	Family	Poaceae
	Hieracium vulgatum	Sub-family	Compositae Taraxacum type
	Calamagrostis purpurea	Family	Poaceae
	Ramischia secunda	Species	
	Equisetum sylvaticum	Genus	
	Antennaria dioica	Sub-family	Compositatae Solidago type

One trial profile has been analysed. This is from Ala-Jalve on the banks of the Teno river, within the birch zone (Fig. 1). The Ala-Jalve site, which has been excavated by RANKAMA (1986), is one of the richest Epi-Neolithic sites in Lapland. It was primarily a summer camp and was used most intensively between 1800 and 800 B.C. This site, and its surroundings, will form one of the major investigation areas for the present project. However, the preliminary pollen diagram presented in Fig. 5 which is from the highest terrace, reflects only its most recent history and illustrates the vegetational development on the site of a fire. The fire horizon is clearly seen in the stratigraphy as a layer of charcoal.

Within this charcoal layer the pollen is highly corroded, as is to be expected, but noticeable are the high values of spores of *Lycopodium annotinum* and the presence of a few pollen grains of *Epilobium*. *Lycopodium* spores are known to be resistent to decay (HAVINGA, 1967) so their abundance may merely reflect the fact that much of the original pollen has been destroyed. *Epilobium*, if this represents *Epilobium angustifolium*, is characteristic of areas which have been burnt. Within these layers *Betula* reaches values of 85-90% P and at the junction between this and the overlying horizon pollen of *Solidago* type is present. The evidence is slight but there is a degree of similarity between this and the species list from the Lapp winter villages further south so this may prove to be a pollen picture for seasonal settlement.

The upper part of the diagram (depth 0-3 cm) shows a more open birch woodland (the proportion of *Betula nana* is also higher in these samples). The proportion of dwarf shrubs is much higher and the presence of both *Pinus* and *Picea* pollen indicates an increase in the amount of long distance transported pollen. The situation resembles that of the present day.

3. Conclusions

This project has just begun but the main lines of approach to be developed are clear:

(1) Use pollen traps to give as complete a pollen picture as possible of each main vegetation belt, taking into account both percentage and influx values;
(2) Compare the pollen trap results for the period 1974 to present with the meteorological records from the same period to establish any possible pollen/climate relationships;
(3) Using the factors obtained from 1 and 2 above analyse long peat profiles for which the time precision is as high as possible, in order to establish the timing and duration of changes in the forest limit which are due to climate. (This represents the area in which the present project interleaves with that of H. HYVÄRINEN);
(4) Use moss samples to obtain pollen pictures of vegetation types influenced by man;
(5) Analyse peat profiles from within and close to known archaeological and historical occupation sites to delimit changes in the forest which are due to human interference and establish the timing and duration of these changes. For this aspect emphasis will be

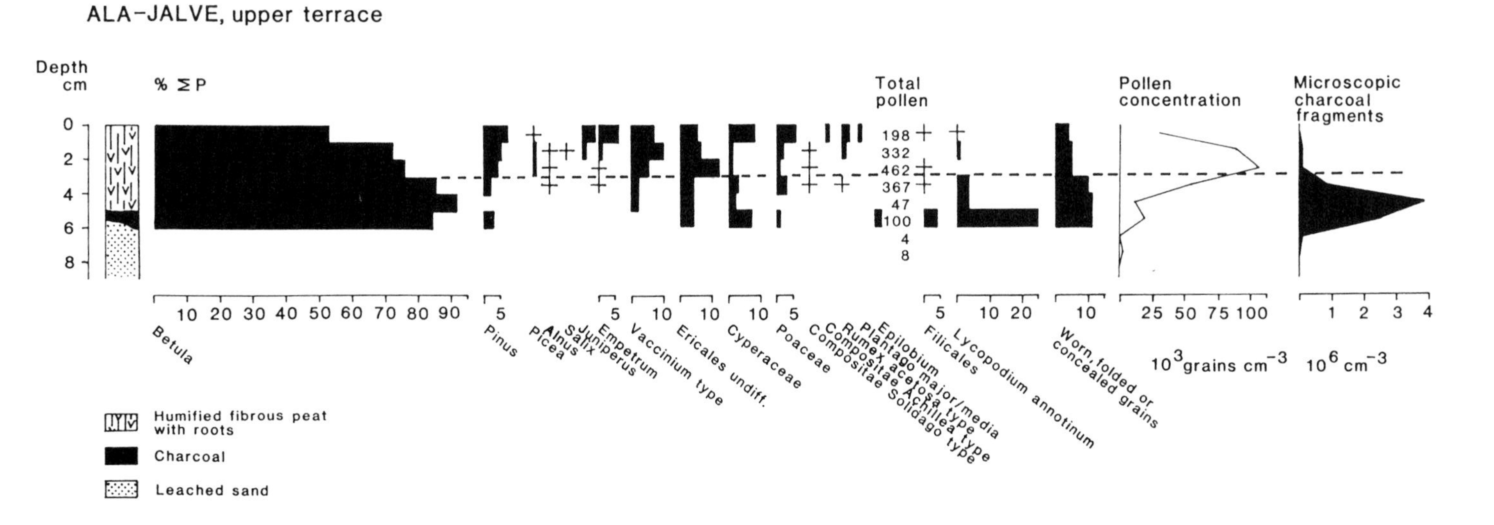

Fig. 5 Preliminary pollen diagram from the epi-neolithic site of Ala-Jalve

placed on two areas, one within the birch woodland (Teno river valley between Ala-Jalve and Utsjoki) and one within the pine forest (Inari and the area to the south and north of the village);

(6) Compare the periods of climatic change as delimited in 3 above with those of human interference as delimited in 5 above.

Acknowledgements

I am most grateful to Christian Carpelan for valuable information and stimulating discussion. My thanks are also due to Tuija Rankama for details of the archaeology of Ala-Jalve, to Kaisu Merenheimo for assistance with the field work, to Raija-Liisa Huttunen for the laboratory preparation of the pollen samples and to Kristiina Karjalainen for help in preparing the diagrams. The project is being financed by the Finnish Academy.

References

CARPELAN, C.; JUNGNER, H. & MEJDAHL, V. (1990): Dating of subrecent Saami winter village site near Inari, Finnish Lapland. Time and environment - A PACT seminar. Abstract volume, 9-10

DAVIS, M. B. (1973): Redeposition of pollen grains in lake sediment. Limnol. Oceanogr. 18, 44-52

DAVIS, M. B. & BRUBAKER, L. B. (1973): Differential sedimentation of pollen grains in lakes. Limnol. Oceanogr. 18, 635-646

HAVINGA, A. J. (1967): Palynology and pollen preservation. Rev. Palaeobot. Palynol. 2, 81-98

HICKS, S. (1985): Modern pollen deposition records from Kuusamo, Finland. I. Seasonal and annual variation. Grana 24, 167-184

HICKS, S. & HYVÄRINEN, V.-P. (1986): Sampling modern pollen deposition by means of "Tauber traps": some considerations. Pollen et Spores 28, 219-242

HYVÄRINEN, H. (1975): Absolute and relative pollen diagrams from Northernmost Fennoscandia. Fennia 142, 23pp

HYVÄRINEN, H. (1976): Flandrian pollen deposition rates and tree-line history in northern Fennoscandia. Boreas 5, 163- 175

KOSKI, V. & TALLQVIST, R. (1978): Tuloksia monivuotisista kukinnan ja siemen sadon määrän mittauksista metsäpuilla. (Results of long-time measurements of the quantity of flowering and seed crop of forest trees). Folia Forest. 364, 1-60

PRENTICE, I. C. (1985): Pollen representation, source area and basin size: toward a unified theory of pollen analysis. Quat. Res. 23, 76-86

RANKAMA, T. (1986): Archaeological research at Utsjoki Ala-Jalve. Helsinki Papers in Archaeology 1, 40 p.

SUOMINEN, J. (1975): Kasvipeitteestä saamelaisten muinaisilla talvikylänpaikoilla. (On the plant cover at sites of ancient Lapp winter villages, Finnish Lapland). Tiedonanataja 79, 92- 94

TAUBER, H. (1965): Differential pollen dispersion and the interpretation of pollen diagrams. Danm. Geol. Unders. IIR 89, 69 p.

TAUBER, H. (1974): A static non-overload pollen collector. New Phytol. 73, 359-369

TAUBER, H. (1977): Investigations of aerial pollen transport in a forested area. Dansk Bot. Ark. 32, 121 p.

Address of author

Dr. S. Hicks, Department of Geology, University of Oulu, Linnanmaa, SF-90570 Oulu, Finland

Holocene pine and birch limits near Kilpisjärvi, Western Finnish Lapland: pollen stratigraphical evidence

Hannu Hyvärinen

Summary

After a brief review of the Holocene tree line history in Northern Fennoscandia, a new pollen site (Masehjavri lake) is described reflecting on the local history of the birch and pine limits near Kilpisjärvi, Western Finnish Lapland. At an altitude of 680 m a.s.l. the site is almost 100 metres above the local birch limit. The pine forest limit lies about 70 km to the south. Changes in pollen composition and concentration suggest that sparse birch woods were present at Masehjavri from about 8000 to 5000 years ago, after which the birch limit has retreated to its present position. This record is compared with a previous record from the nearby site of Mukkavaara situated, at 535 m a.s.l., below the present birch limit. Pine woods were present at Mukkavaara during the time when the birch limit was higher than today, but pine never reached the altitude of Masehjavri. The retreat of pine from the area approximately 5000-4000 years ago seems to parallel the drop of the birch limit.

Zusammenfassung

Der Artikel gibt eine kurze Übersicht über die postglaziale Entwicklung der nordfennoskandischen Waldgrenzen im Lichte früherer pollenanalytischer Untersuchungen und diskutiert gleichzeitig die lokale Geschichte der Birken- und Kieferngrenzen bei Kilpisjärvi im westlichen Finnisch-Lappland aufgrund eines neuen Pollenprofils von Masehjavri. Der See Masehjavri liegt mit einer Höhe von 680 m im baumlosen Bereich; die Birkengrenze liegt ca. 100 Meter tiefer, und die heutige Nordgrenze des Kieferngebietes befindet sich über 70 km weiter im Süden. Frühere Pollenergebnisse und Makrorestfunde zeigen, daß die Kiefer zwischen 7500 und 4000 J.v.h. in der Umgebung von Kilpisjärvi gewachsen ist. Nach den neuen Ergebnissen gab es zu dieser Zeit Birkenbestände im Höhenbereich von Masehjavri, jedoch keine Kiefern. Schon im früheren und mittleren Postglazial ist also eine Birkenstufe über der maximalen Kieferngrenze nachweisbar, und diese beiden Grenzen haben sich im Verlauf der letzten 5000-4000 Jahre parallel zueinander zurückgezogen.

1. Previous studies

The Holocene history of the pine limit in Northern Fennoscandia has been discussed in several contexts on the basis of both pollen and megafossil data collected in the 1970s and 80s (HYVÄRINEN, 1975, 1976, 1985; ERONEN, 1979; ERONEN & HYVÄRINEN, 1982; ERONEN & HUTTUNEN, 1987). The pollen evidence is mainly based on pollen influx studies of lake sediment cores from the outer parts of the present pine region and from the mountain birch region (Fig. 1). While the abundant finds of pine megafossils give a good picture of the former distribution of pine and provide concrete evidence for the local presence of pine trees at a certain place and time, pollen data, especially pollen influx data, can provide evidence for the absence of pine woods from an area and for the changing densities of pine populations in the landscape.

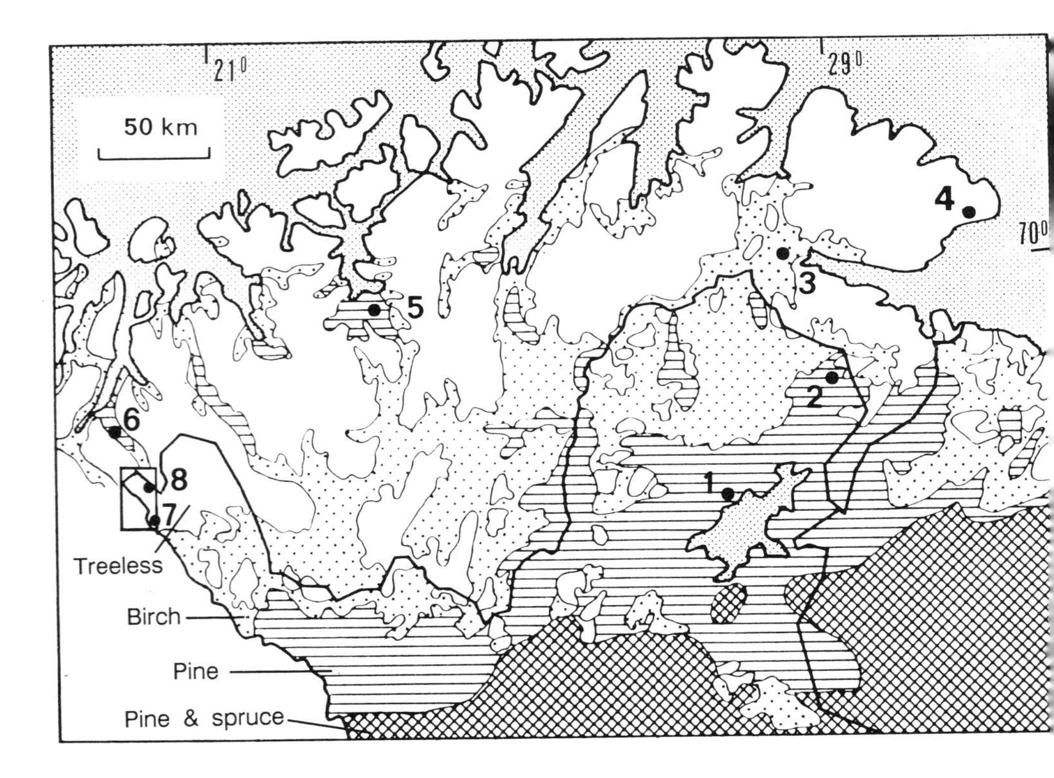

Fig. 1 Map of Northern Fennoscandia showing the main vegetation regions and the location of Holocene pollen diagrams from lake sediments. (1) Akuvaara; (2) Suovalampi; (3) Bruvatnet; (4) Domsvatnet (HYVÄRINEN, 1975, 1976); (5) Alta (HYVÄRINEN; 1985); (6) Råttuvarri; (7) Mukkavaara (ERONEN & HYVÄRINEN, 1982); (8) Masehjavri (present paper). The area of Kilpisjärvi (Fig. 2) is shown by rectangles

Pollen-analytical studies in Northern Fennoscandia were initiated by AARIO, who first demonstrated the potential of absolute pollen counting as a method capable of elucidating the history of the northern forest limits (AARIO, 1940). However, in AARIO's time it was not possible to obtain adequate sample series with good time control (control of matrix sedimentation rates) for the generation of meaningful stratigraphical pollen influx records. When pollen studies in the forest limit area were resumed in the 1970s (HYVÄRINEN, 1975), cores were taken from small, closed lake basins, which could be expected to contain organic sediments accumulated at an even rate and suitable for radiocarbon dating. Comparison of core-top values showed that recent or subrecent pine pollen accumulation rates were quite different at sites within and outside the present pine limit, a feature also evident from subsequent investigations of modern pollen rain (HICKS, 1977; PRENTICE, 1978). Cores from outside the present pine limit yielded high fossil rates, indicating that pine was earlier present at these sites, and a retreat of the pine limit is indicated by declining rates towards the present. No corresponding change was observed at sites which today are still within the pine forest.

Briefly, the available pollen evidence, in agreement with the megafossil evidence, suggests that pine spread to northernmost Fennoscandia between 8000 and 7000 yr B.P., replacing the initial, Early Holocene, birch forests practically everywhere at lower elevations. After that a period of relative stability followed, lasting until 4500-5000 yr B.P. when a regional retreat of the pine limit started, leading gradually to the present situation. Isolated islands of pine survived in sheltered valleys and fjord heads as remnants of the earlier, more extensive pine region. The megafossil evidence from the Swedish Scandes indicates a more or less parallel drop in the altitudinal pine limit since about 4000 years ago (KULLMAN, 1990). The retreat of the pine limit in the north as well as farther south in the Scandes is generally attributed to long-term climatic deterioration in Late Holocene times.

It is less clear to what extent the initial birch forests were replaced by pine and what the position of the Early and Middle Holocene birch limit was after the pine invasion. The origin of the Fennoscandian mountain birch region has been earlier discussed mainly with reference to conditions in the Southern Scandes (AAS & FAARLUND, 1988, KULLMAN, 1990). The mountain birch woods are generally regarded as an oceanic feature and, in accordance with NORDHAGEN (1933), many Scandinavian authors have postulated that the birch belt is a Late Holocene development associated with a trend of increasing climatic oceanity after a more continental Early Holocene. KULLMAN finds no indications of a birch belt in the Swedish Scandes until after about 6000 yr B.P.; pine seems to have formed the Early Holocene tree line there, and the birch belt was only established later on when "the sparse pine tree limit ecotone started regression and was replaced by a growing fringe of subalpine deciduous forest" (KULLMAN, 1990). On the other hand, there is positive megafossil evidence from the more western (and oceanic) parts of the Scandes in Southcentral Norway for a high birch limit above the Early to Middle Holocene pine limit (AAS & FAARLUND, 1988).

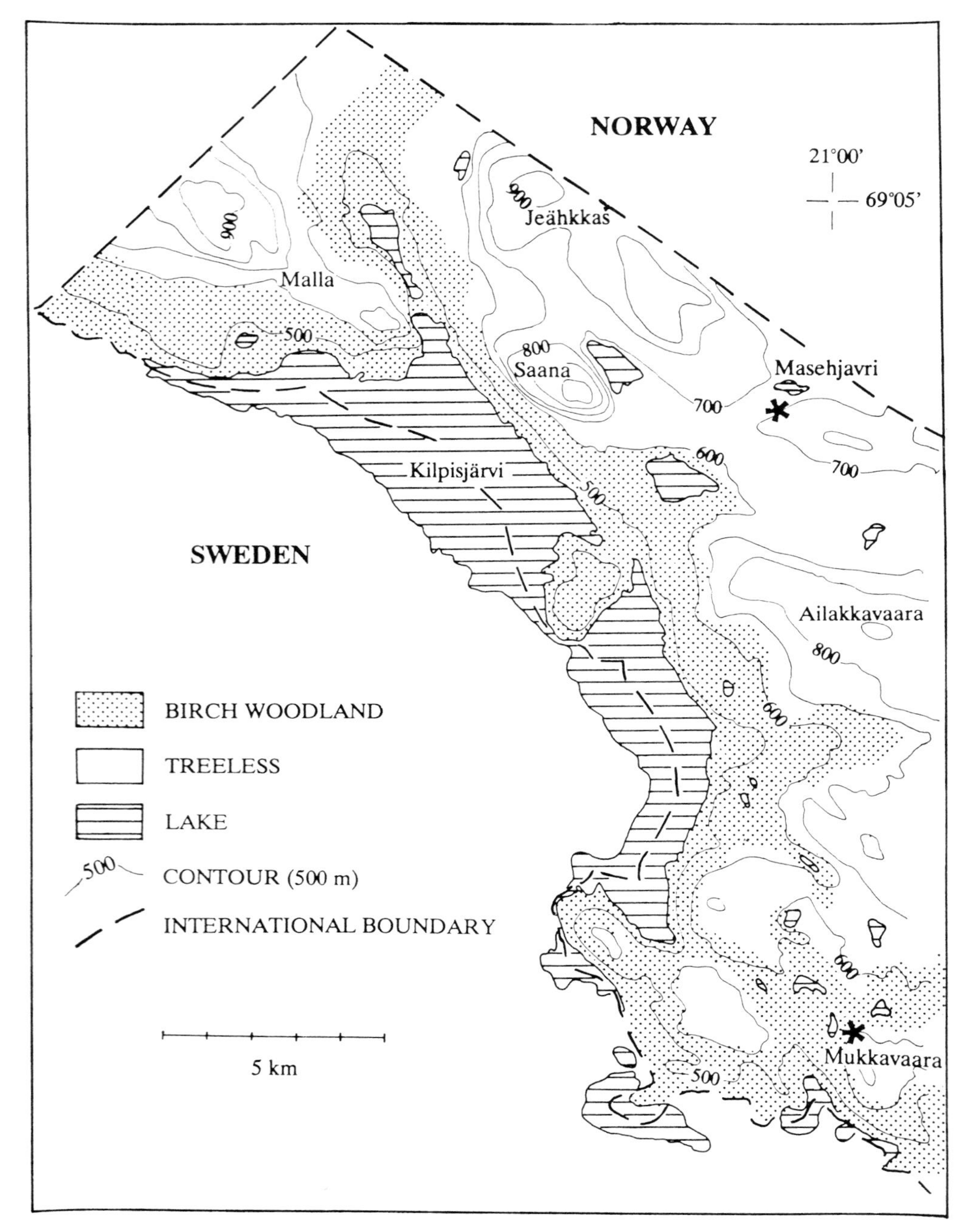

Fig. 2 Map of the Kilpisjärvi area. The pollen sites of Masehjavri and Mukkavaara are marked by asterisk

The evidence available from Northern Fennoscandia is sparse. A pollen record from lake sediments on Varanger Peninsula (HYVÄRINEN, 1976) was used to demonstrate that birch has always formed the outermost woods there. According to the early pollen studies by AARIO (1943), this was also the situation in Petsamo, an area bordering on the Kola Peninsula. However, informative pollen records from the barren areas outside the modern birch limit are rare, and no relevant megafossil evidence is available either. The new pollen data presented here sheds light on the birch limit history in Western Finnish Lapland; at the same time they illustrate the use of stratigraphical pollen evidence in tree line studies in general.

2. Pollen evidence from the Kilpisjärvi area

The map in Fig. 2 shows the distribution of the birch woodland in the Kilpisjärvi area and the location of the two pollen sites discussed here. Mukkavaara is an old site described by ERONEN & HYVÄRINEN (1982). Masehjavri was cored at the same time as Mukkavaara, but has not been published previously.

Masehjavri (69°03'N, 20°59'E) is a lake situated at 680-690 m a.s.l., almost 100 m above the local birch limit. It measures about 900 by 350 m and has a maximum depth of approximately 8 m. Sample cores were taken in 1976 from a depth of 6.2 m. Stratigraphy: 0-143 cm, greenish brown gyttja; 143-160 cm, grayish green silty gyttja grading into gray silt; 160-170 cm, sandy silt and sand. The radiocarbon dates (Fig. 3) suggest an even rate of sedimentation, but appear consistently to be too old. The topmost sediment (0-10 cm) was dated at 1680±110 yr B.P., although there is no indication of anything missing from the top, and also the rest of the dates down the core are between 500 and 1000 years older than could be expected on the basis of the deglaciation history and pollen stratigraphy. The lake basin lies on the Precambrian crystalline basement, but the hill slopes immediately to the west (Jeähkkas and Saana) consist of Cambro-Silurian sedimentary rocks including dolomites; hence the dates from Masehjavri may be influenced by the hard water effect.

The pollen diagrams (Fig. 4, 5) are abbreviated APF (absolute pollen frequency per cm^3 of fresh sediment) diagrams. As the sediments are homogeneous gyttjas unlikely to show large changes in the rate of deposition, the APF variations are believed to represent true variations in the pollen input to the lakes. However, in the topmost part (approx. 1 m) of the Mukkavaara core, consisting of a loose gyttja with abundant peat detritus, the rate of matrix deposition seems to be higher than further down (ERONEN & HYVÄRINEN, 1982), and the consistently low APF values for all taxa in this part of the core are probably unrepresentative.

The diagrams can be readily correlated on the basis of the regional features in the pollen stratigraphy, mainly birch, pine, alder, and spruce curves: a basal birch dominance is followed by maxima of pine and alder, and the uppermost zone shows a continuous spruce curve associated with declining alder. The spruce pollen is totally derived from distant sources, but provides a useful stratigraphic marker. The spruce pollen limit has an age of 3000-3500 yr B.P., and the rise of pine can be dated at between 7000 and 7500 yr B.P. (ERONEN & HYVÄRINEN, 1982; HYVÄRINEN, 1985).

Comparison of the local pollen zones reveals marked differences between the sites. In both cases the stratigraphy opens with a birch-Ericales zone with low APF (Ma 1, Mu 1). The subsequent rise in birch APF represents the establishment of birch woodland (Ma 2, Mu 2). The high birch values are initially associated with abundant lycopods, ferns, and Poaceae (subz. 2a) followed by a rise in juniper (2b). Today juniper is a common and abundant component of the groundcover of the mountain birch woods, and juniper pollen is universal in modern pollen spectra from this region (PRENTICE, 1978); hence the birch-juniper assemblage can be taken to represent a birch woodland of the present type. At Mukkavaara the birch woodland is soon replaced by woods with an admixture of pine (Mu 3); the change in vegetation is indicated by the drop in birch and juniper following the rise in pine.

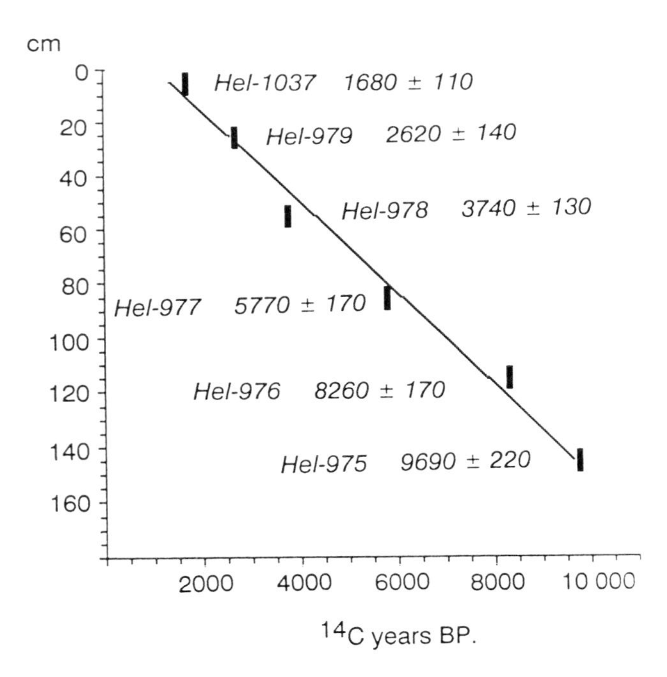

Fig. 3 Time/depth distribution and trend of radiocarbon dates from the Masehjavri core. The dates are considered to be too old by between 500 and 1000 years throughout

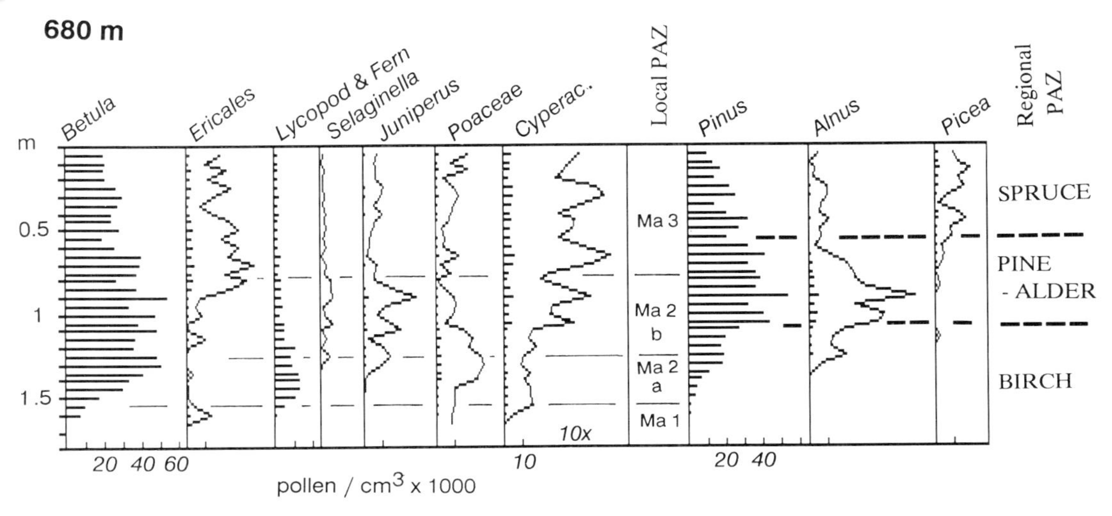

Fig. 4 Abridged pollen diagram of Masehjavri

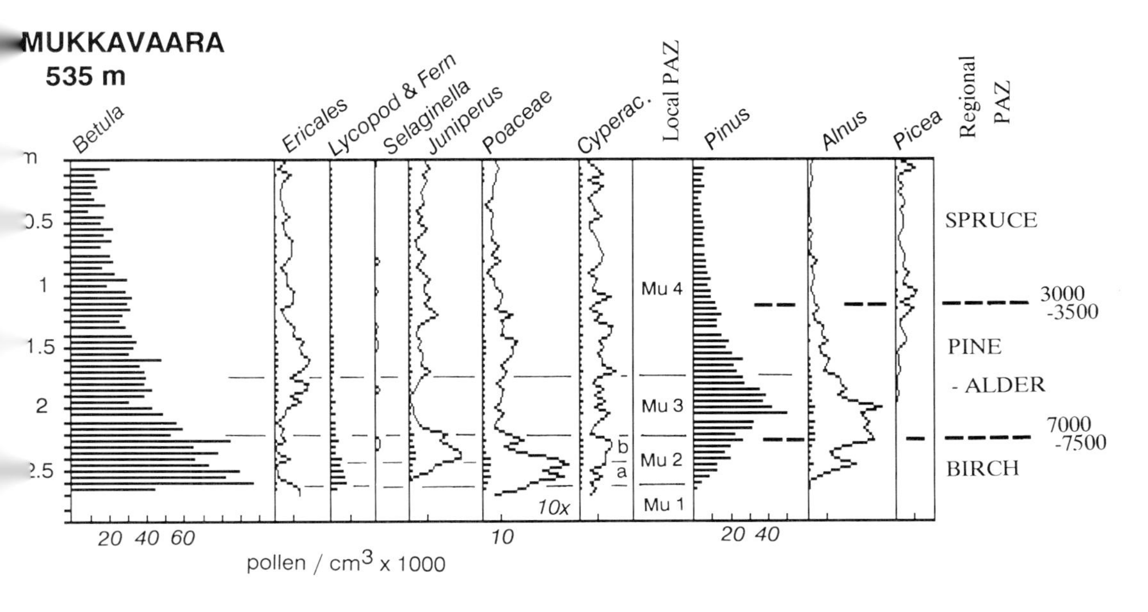

Fig. 5 Abridged pollen diagram of Mukkavaara

The local presence of pine is confirmed by megafossil finds from lakes nearby (ERONEN & HYVÄRINEN, 1982; ERONEN & HUTTUNEN, 1987). At Masehjavri, on the other hand, no corresponding change is observed; the rise of pine causes no decline in birch and juniper, and the local vegetation must have remained birch woodland. The uppermost zone at Masehjavri (Ma 3) opens with a rise in Ericales and Cyperaceae and a decline in birch and juniper, reflecting the establishment of the present type of barren dwarf-shrub heath. The declining pine values in both diagrams result from the regional retreat of the pine limit, and at Mukkavaara birch woodland is re-established (reappearance of juniper). As pointed out above, the uppermost APF trends (general decline) at Mukkavaara may in part reflect an increased rate of matrix sedimentation.

3. Discussion

In conclusion, the results from Kilpisjärvi show that the Early and Middle Holocene birch limit in this area was higher than today and clearly above the maximum pine limit, and that the pine and birch limits experienced a more or less parallel retreat in Late Holocene times. The available data do not allow very exact conclusions about the former extent of the birch region. Horizontally, the birch region in Northern Fennoscandia has probably never been wider than today, but this may just be a reflection of the flat topography of the areas now occupied by birch. At Kilpisjärvi, for instance, a vertical drop in the birch limit from at least 700 metres (Masehjavri) to the present 600 metres can be demonstrated. This meant a horizontal change of a few or a few scores of kilometres. The vertical drop in the pine limit was of the same order of magnitude, from about 550 to 400 metres, but horizontally this brought the pine limit back and thus extended the birch region by between 50 and 100 kilometres.

The data from the Swedish Scandes suggest that there were significant regional differences in the history of the tree line forests, reflecting in all probability regional differences in climate. In the more continental areas, such as the eastern slopes of the Southern Scandes, a well-defined birch belt seems to have developed only during the latter half of the Holocene. Even today the birch belt is poorly developed or totally absent from hill massifs rising above the coniferous forest limit in central parts of Finnish Lapland. However, the results from Kilpisjärvi together with similar results from the Varanger Peninsula and from Southern Norway show that in much of Fennoscandia a belt of birch woodland against the barren areas has been a permanent feature of vegetation throughout the Holocene.

References

AARIO, L. (1940): Waldgrenzen und subrezente Pollenspektren in Petsamo Lappland. Ann. Acad. Sci. Fenn. A 54/8, 120 p.

AARIO, L. (1943): Über die Wald- und Klimaentwicklung an der lappländischen Eismeerküste in Petsamo. Ann. Bot. Soc. Zool. Bot. Fenn. "Vanamo" 19/1, 158 p.

AAS, B. & FAARLUND, T. (1988): Postglasiale skoggrenser i sentrale sørnorske fjelltrakter. ^{14}C-datering av subfossile furu- og bjørkrester. (Postglacial forest limits in Central South Norwegian mountains. Radiocarbon datings of subfossil pine and birch specimens). Norsk Geogr. Tidsskr. 42, 25-61

ERONEN, M. (1979): The retreat of pine forest in Finnish Lapland since the Holocene climatic optimum: a general discussion with radiocarbon evidence from subfossil pines. Fennia 157, 93-114

ERONEN, M. & HYVÄRINEN, H. (1982): Subfossil pine dates and pollen diagrams from Northern Fennoscandia. Geol. Fören. Stockh. Förh. 103, 437-445

ERONEN, M. & HUTTUNEN, P. (1987): Radiocarbon dated subfossil pines from Finnish Lapland. Geogr. Ann. 69 A, 297-304

HICKS, S. (1977): Modern pollen rain in Finnish Lapland investigated by analysis of surface moss samples. New Phytol. 78, 715-734

HYVÄRINEN, H. (1975): Absolute and relative pollen diagrams from northernmost Fennoscandia. Fennia 142, 1-23

HYVÄRINEN, H. (1976): Flandrian pollen deposition rates and tree-line history in Northern Fennoscandia. Boreas 5, 163-175

HYVÄRINEN, H. (1985): Holocene pollen history of the Alta area, an isolated pine forest north of the general pine forest region in Fennoscandia. Ecologia Mediterranea 11, 69-71

KULLMAN, L. (1980): Radiocarbon dating of subfossil Scots pine (*Pinus sylvestris* L.) in the Southern Swedish Scandes. Boreas 9, 101-106

KULLMAN, L. (1990): Dynamics of altitudinal tree-limits in Sweden: a review. Norsk Geogr. Tidsskr. 44, 103-116

NORDHAGEN, R. (1933): De senkvartaere klimavekslinger i Nordeuropa og deres betydning for kulturforskningen. Inst. for sammenlingn. kulturforskn. Ser. A: Forelesninger 12, 246 p.

PRENTICE, C. (1978): Modern pollen spectra from lake sediments in Finland and Finnmark, North Norway. Boreas 7, 131-153

Address of the author:

Dr. H. Hyvärinen, Department of Geology, Division of Geology and Palaeontology, University of Helsinki, Snellmaninkatu 5, SF-00170 Helsinki, Finland

Pine megafossils as indicators of Holocene climatic changes in Fennoscandia

Matti Eronen & Pertti Huttunen

Summary

A large number of subfossil pine trunks have been preserved in an area of Fennoscandia lying beyond the present pine limit, as a reminder of the times when pines used to occur higher on the slopes and over a more extensive area of the north than they do today. Numerous ^{14}C dates have been obtained for pine megafossils, of which more than 280 are discussed in this paper, mostly collected from various publications. The temporal distribution of the dates is examined with respect to individual areas. In Southern Fennoscandia the pine limit reached its highest levels in the Early Holocene, while the pine forests are estimated to have reached their maximum extent in the north during the Middle Holocene. There were thus differences with respect to climatic changes between the regions of Fennoscandia during the Holocene. Pine growth and reproduction are dependent on summer temperatures in particular, i.e. conclusions on temperature changes can be drawn from the pine megafossil material mainly.

Zusammenfassung

Subfossile Kiefernstämme sind in Fennoskandien in großer Zahl oberhalb der Kieferngrenze erhalten, als Relikte einer Zeit, in der die Kiefern an den Berghängen höher hinaufreichten und in nördlicheren Regionen wuchsen als heutzutage. Viele der Kiefernmegafossilien wurden mit Hilfe der ^{14}C-Methode datiert. Im Rahmen dieser Untersuchung konnte auf über 280 Datierungen, zumeist aus Veröffentlichungen, zurückgegriffen werden. Anhand der regionalen und zeitlichen Verteilung der ^{14}C-Daten konnte für das südliche Fennoskandien das frühe Holozän als Periode der größten Ausdehnung der Kiefernwälder ausgegliedert werden, für die nördlicheren Bereiche dagegen das mittlere Holozän. Hieraus wird abgeleitet, daß die holozäne Klimaentwicklung in Fennoskandien regional differenziert verlief. Da das Wachstum und die Ausbreitung der Kiefer inbesondere von den Sommertemperaturen abhängig ist, kann man aufgrund der Kiefernmegafossilien Rückschlüsse auf Klimaveränderungen ziehen.

1. Introduction

Northern Fennoscandia is characterized by a distinct vegetational zonality, indicated by the gradual transition from boreal mixed coniferous forests first to pine forests, then to alpine birch forests, and finally to bare fell tops (see the preceding introductory section). This zonality is not a static feature, of course, but shaped by climatic changes. Ecotonic boundaries changed throughout the Holocene, and even observations made in the last few decades indicate changes in forest boundaries (HUSTICH, 1958; KULLMAN & HÖGBERG, 1989; KULLMAN & ENGELMARK, 1990). It is this sensitivity of a marginal zone to various environmental changes which makes Northern Fennoscandia a highly interesting area for investigations into Postglacial vegetational development.

Birch and pine spread to the northern parts of Fennoscandia during the Early Holocene, the former following close on the continental ice sheet as it retreated from Finnish Lapland by 9000 yr B.P. and melted completely in Sweden by 8500 yr B.P. (LUNDQVIST, 1986). The climate may have been quite warm during the Boreal chronozone, around 8000-9000 yr B.P., as suggested by the rapid melting of the continental ice sheet and the considerable decrease in the alpine glaciers of Scandinavia at this time (MANGERUD, 1990; NESJE & DAHL, 1991; NESJE et al., 1991). The forests advanced northwards so slowly, however, that this warm period does not stand out from the general trend in the vegetational history of Northern Fennoscandia, where the rise in temperature is indicated in palaeobotanical material by the emergence of Scots pine (*Pinus sylvestris*, L.), although with a probable delay of approx. 1000 years, attributable to the migratory history of this species. It nevertheless seems that there have also been differences between various regions of Fennoscandia in terms of climatic history.

2. Pine megafossils

Once pine had become established in Fennoscandia it became far more widespread than at present, growing on the fells above the present pine boundary and covering a wider area in the north than it does today. Thousands of pine trunks have been found in mires and small lakes in the region beyond the present *Pinus* limit where they had been preserved for several thousand years. Identification to species is easy, as *Pinus sylvestris* is the only large arboreal species to have occurred as far north since the last glaciation.

The earliest reports which briefly discuss trunks found beyond the present pine limit date from the nineteenth century (ERONEN, 1979), but these subfossils achieved true significance for research purposes with the introduction of the radiocarbon dating method only. This method was first applied to pine megafossils in Fennoscandia at the Laboratory of Radiocarbon Dating in Stockholm in the 1950s (LUNDQVIST, 1959).

All the ^{14}C dates published for megafossil pines in Fennoscandia, together with 18 dates obtained by BARTHOLIN for dendrochronological purposes which have not been published as such (see BARTHOLIN, 1987), are collected in Fig. 1. Five areas in which such finds have been made are separated in Fig. 2, together with the numbers of megafossils dated from each.

The age of the dated pine trunks and altitudes of the sampling points (a.s.l.) are also shown in Fig. 1. The resulting pattern is quite incoherent, however, due to the fact that the altitude of the pine limit changes considerably from the fells of Southern Norway, where pine grows up to 900 m, northeastwards across Fennoscandia, so that there are places beside the fjords of Northern Norway where pine is not found growing naturally at all, even at sea level.

A significant feature in Fig. 1 is the low number of dates recorded between 2500 and 2000 yr B.P. A number of investigators have pointed out that the climate deteriorated in Europe around 2500 yr B.P. (see e.g. LAMB, 1982; GROVE, 1988) and it is quite possible that this gap in the dates is caused by a contemporaneous retreat of the pine limit, a conclusion which would require confirmation through further research. More information about this question will be obtained when the dendrochronological master curve of pine which at present is in the process of being compiled has been extended sufficiently far back in time (ERONEN et al., 1991).

The material represented in Fig. 1 has been collected from the pine tree limit zone and areas beyond it in different parts of Fennoscandia over a considerable period of time, and it is naturally the case that the number of subfossil pines found has greatly exceeded the available ^{14}C dating capacity. The samples mainly have been selected at random, however, so in this respect the material provides a relatively homogeneous base for describing the age distribution. When collecting the extensive material required for the present dendrochronological survey it has sometimes been necessary to reject badly decayed trunks and even undamaged stumps if the set of growth rings available was not of sufficient length or quality. The "floating chronologies" already involve a relatively high number of dendrochronologically dated subfossil trees, which are of course not included in Fig. 1 (cf. BARTHOLIN, 1987; ERONEN et al., 1991).

3. Areal variations in changes of the pine limit

The preciseness of the data obtained from the material varies, the dates from Norway, for example, partly lacking accurate data on the altitude difference between the sampling site and the present limit of pine. For this SELSING & WISHMAN (1984) report a value of 65-115 m in their areas of investigation on the Southwest Norwegian mountains, while the authors report that the present pine limit there lies at 630-930 m a.s.l. According to AAS &

FAARLUND (1988) subfossil pines can be found in the Hardangervidda and Jotunheimen areas in Central Southern Norway at altitudes 1250-1300 m a.s.l. The ^{14}C dates show that the pine limit was at such high levels, 250-300 m above the present one, between 8000 and 7000 yr B.P.

It is impossible to give an exact value for such vertical differences in Finnish Lapland as the role of horizontal transitions has generally been more significant, owing to the relatively low relief of the terrain. The highest occurrences of pine subfossils in the north-western corner of the country, at Enontekiö in Lapland, are found at altitudes of 505-560 m a.s.l., while the last pine forests located within the borders of Finland lie at an altitude of approx. 400 m a.s.l. There is a distance of 60-70 km between these sites and the places of discovery of the above pine subfossils, although the pine forest islet of Lyngenfjord, lying beyond the watershed in Norway, begins a horizontal distance of slightly over 20 km away. It is nevertheless impossible to provide an exact picture of the transition of the pine limit in Finnish Lapland merely by indicating its decline in a vertical direction.

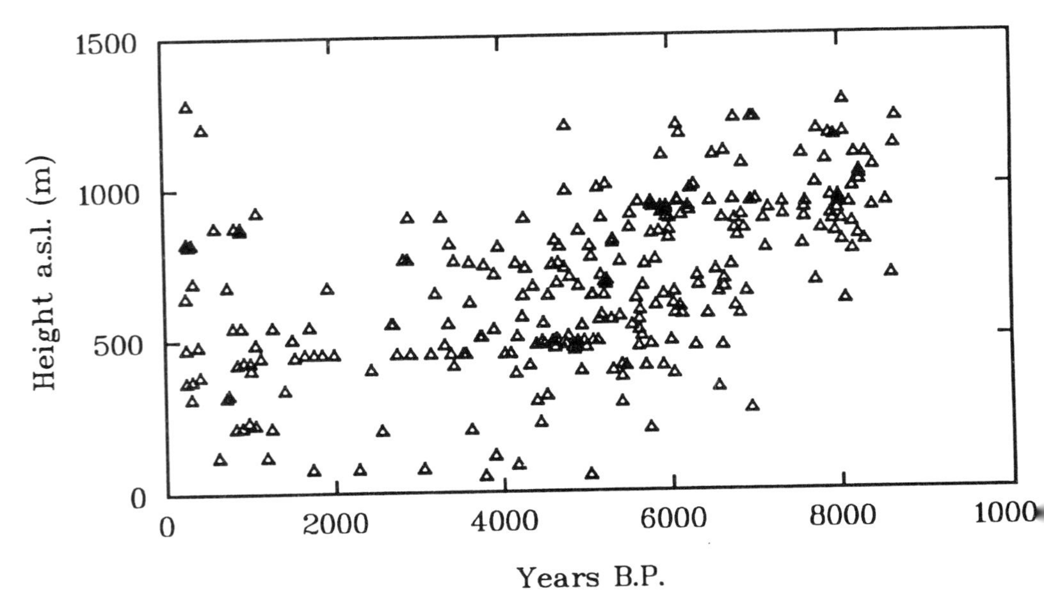

Fig. 1 ^{14}C dates for subfossil pines collected from Fennoscandia. The vertical scale shows the altitudes of the sites a.s.l. and the horizontal scale their ages. Sources: AAS & FAARLUND (1988), BARTHOLIN (unpublished dates), ERONEN & HYVÄRINEN (1982), ERONEN & HUTTUNEN (1987), HAFSTEN (1981), KARLÉN (1976), KULLMAN (1987, 1988, 1990a), KULLMAN & ENGELMARK (1990), MOE (1979), SELSING & WISHMAN (1984)

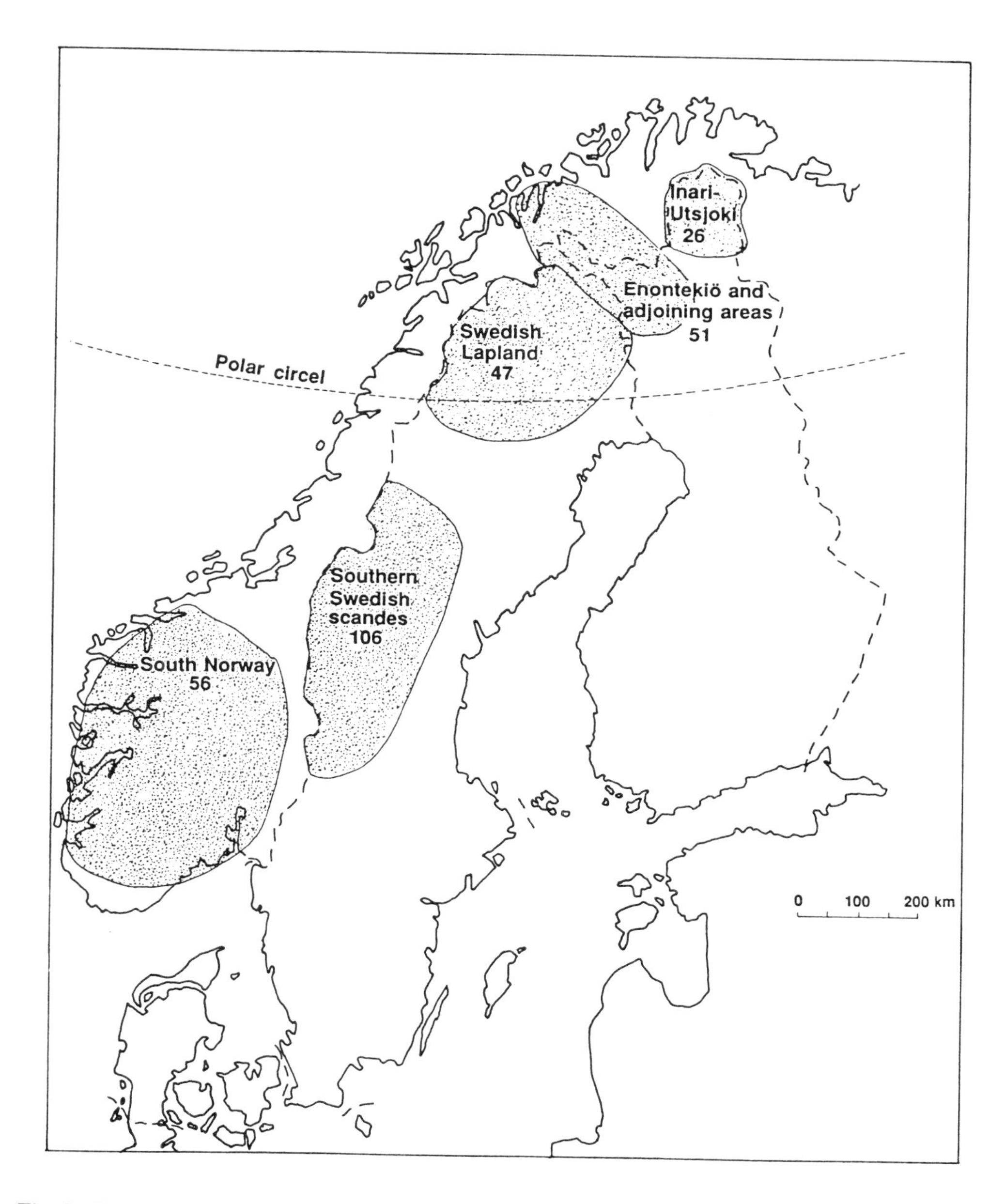

Fig. 2 Five regions of Fennoscandia of interest for the examination of pine megafossil material and numbers of ^{14}C-dates obtained from them

As Swedish Lapland and the Scandes mountains are characterized by a much steeper topography than Finnish Lapland, it has been possible to give the altitude of the pine samples dated in Sweden in metres relative to the present pine limit. The figures may be somewhat vague, as the tree limit is a relatively diffuse point of reference. It is difficult to define exactly and can vary locally and temporarily for a variety of causes. Dates recorded for subfossil pines found above the present limit in Swedish Lapland and the Southern Swedish Scandes are presented in Fig. 3.

The earliest ^{14}C dates in Fig. 3 are markedly older in the Southern Scandes than in the north, which is explained by the migratory history of pine. The limit for the species had already declined in the Southern Scandes by the time it had reached its highest growing sites in the north, in 7000-4000 yr B.P. The greatest altitude difference between the highest subfossil pines and the present tree limit is approx. 250 m in the north and approx. 200 m in the south, and even the average altitude difference between them is greater in the north than in the south. Pines were growing in the north over 125 m above the present limit in 6000-4000 yr B.P., while a less marked altitude difference was observed in the Southern Scandes, apart from one exception shown in Fig. 3. This may provide some information on regional differences in climatic development, as the pine limit in the south at approx. 8000 yr B.P. was at about the same altitude above the present pine tree limit as at the maximum stage in the north.

Land uplift has had some effect on Postglacial climate and the tree limit in Fennoscandia. It can be inferred from the present rate of uplift and from the ancient raised shorelines that the areas featuring pine limits will have been raised several tens of metres since the invasion of pine (Ekman, 1987; Eronen, 1988). However, uplift only accounts for a minor proportion of the long term decline in the pine limit and cannot explain the regional differences.

The dependence of the growth of forest limit pines on summer temperatures was convincingly demonstrated by Hustich (1948) and later confirmed in many investigations (see Eronen et al., 1991). Therefore it is quite certain that changes in the amount of incoming solar radiation caused by astronomic factors were involved in the movement of the pine limit. In the course of the precession cycle, the earth came nearest to the sun in the northern hemisphere around 11,000 yr B.P. during the summer, while today it reaches that perihelic position of its elliptic orbit in the north during winter. In addition, the angle of the earth's orbit with respect to its axis of rotation has decreased, which has also reduced radiation at high latitudes during the northern summer (Imbrie & Imbrie, 1979; COHMAP, 1988). Thus the intensity of the solar radiation reaching the northern hemisphere in summer has decreased during the last 11,000 years. The amount of radiation during summertime was markedly higher than today in 9000-6000 yr B.P. and may have facilitated the survival of pine in the north.

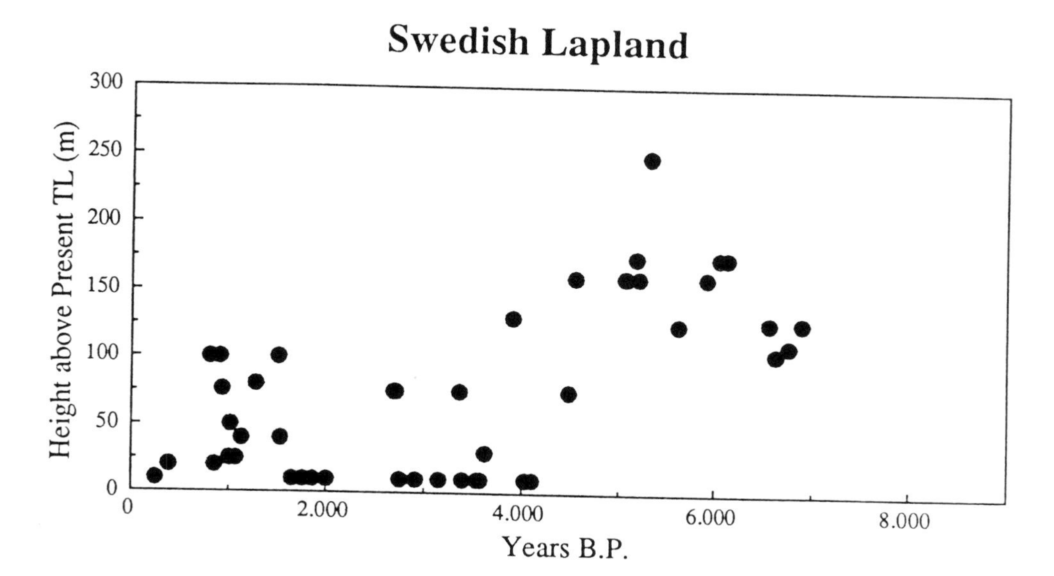

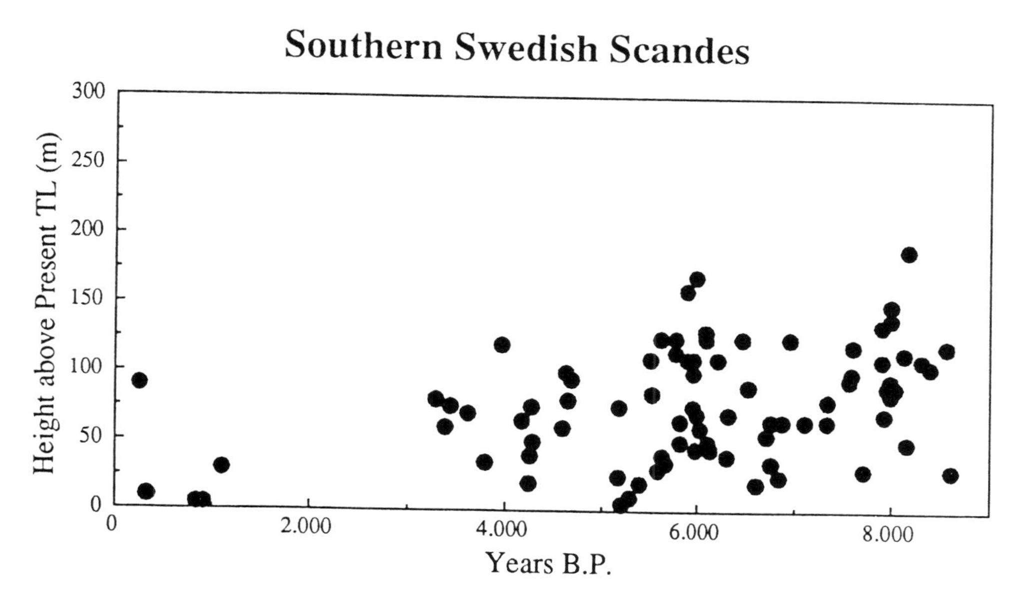

Fig. 3 Dates obtained for subfossil pines found above the present pine limit in Swedish Lapland and the Southern Swedish Scandes. The vertical scale indicates the altitudes of the sites above the present pine limit. Sources: BARTHOLIN (unpublished dates), KARLÉN (1976), KULLMAN (1987, 1988, 1990a), KULLMAN & ENGELMARK (1990)

The significant role of summer temperatures with respect to the spread of pine becomes particularly evident when one looks at its sexual regeneration, which is dependent on climatic conditions during three growth seasons: the first summer, when the buds emerge, the second summer, when the flowers develop, and the third summer, when the seeds mature. Low summer temperatures reduce flowering, but they are most critical of all during the maturing year, as this requires a temperature sum of 950 degree days during the growth period, although partial maturing will still occur at 700-500 degree days, a level sometimes reached in the forest limit zone (NUMMINEN, 1989). However, even minor long-term drops in temperature are sufficient to prevent sexual reproduction and will thus force regression of the pine limit.

On the other hand, the observations reported by KULLMAN & HÖGBERG (1989) show that a cold-induced dieback of mature forest limit pines can occur under certain conditions as a consequence of exceptionally severe winter times.

4. General pattern of changes in the pine limit in Fennoscandia

Fig. 4 gives an outline of the changes which have taken place in the pine limit in four areas in Fennoscandia during the Holocene.

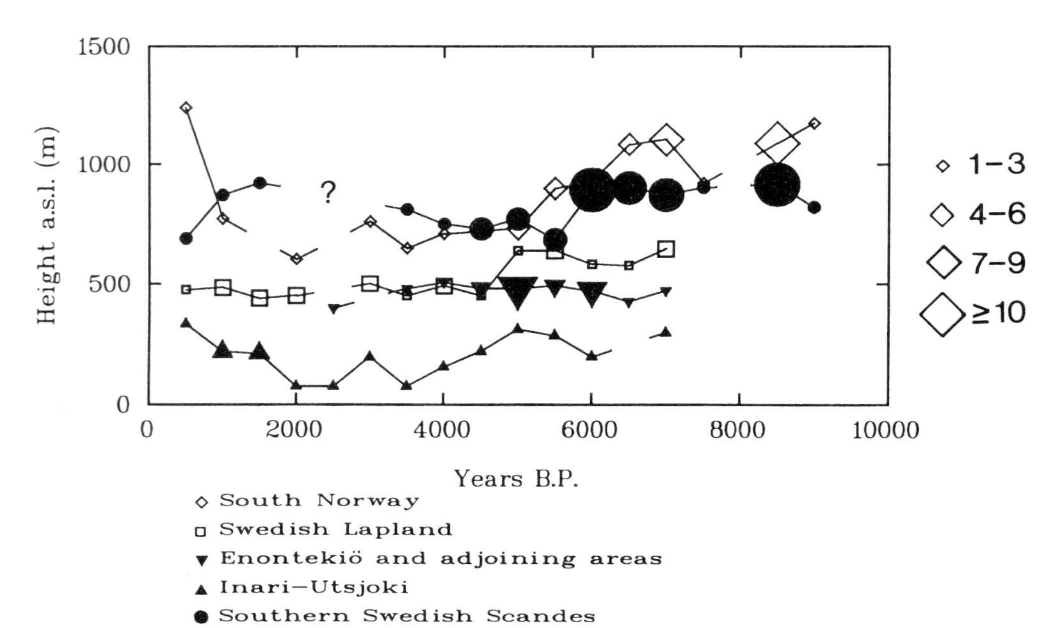

Fig. 4 General pattern of changes in the pine limit in the five regions of Fennoscandia examined (see Fig. 2). References the same as in Fig. 1

The ^{14}C dates were first divided into groups, each covering a period of 500 years, after which a median altitude a.s.l. was calculated for each. The sizes of the symbols used in Fig. 4 indicate the numbers of occurrences in each group in terms of selected size categories.

The procedure employed when compiling the diagram in Fig. 4 eliminates possible short-term variations in the altitude of the pine limit, which according to KARLÉN (1976) may have occurred quite frequently during the Holocene (cf. KULLMAN & ENGELMARK, 1990). The aim here is not to look at the details, however, but rather to point out the general differences between regional changes and the method of analysis employed here was chosen with this in mind. In addition, it is quite difficult to provide a reliable account of minor oscillations in the pine limit which are hampered by the diffuse character of the boundary. Although the ^{14}C-dates number over 280, they are still insufficient to provide a reliable indication of short-term fluctuations in the forest limit, nor have pollen analyses proved effective in this respect (HYVÄRINEN, 1975, 1976, this volume). Occasional local fluctuations are insignificant for palaeoclimatological research, as only extensive, distinctive changes can be linked with macroclimatic variations. An extensive short-term oscillation in the pine limit in Scotland has recently been described by GEAR & HUNTLEY (1991), but the present material from Fennoscandia is not indicative of any similar sudden large-scale changes. The pine annual ring chronology covering several thousands of years which is currently being compiled for Fennoscandia will undoubtedly be of major importance for investigations into short-term climatic changes.

Fig. 4 once again indicates that pines used to grow high up in the Scandinavian Mountains during the Early Holocene, the earliest occurrences of this highest pine limit (approx. 9000-8000 yr B.P.) being recorded in Southern Norway. Dates from the Southern Scandes indicate that considerable numbers of megafossils exist in the age groups 8000-8500 and 5500-6000 yr B.P., which may be a sign of elevation of the tree limit at these times (cp. KULLMAN, 1990b). There would seem to be a temporary drop and a new rise between 8000 and 7000 yr B.P., followed by a marked drop in 7000-6000 yr B.P. In the case of Swedish Lapland and of Enontekiö in Finland, however, the pine limit seems to have remained at a relatively fixed level during the interval 7000-5000 yr B.P. It was situated somewhat lower in Enontekiö than in Swedish Lapland at first, but a drop which occurred in the latter area in 5000 yr B.P. brought it to the level of the former, especially since there was no such simultaneous drop in Enontekiö. There are a considerable number of dates from Enontekiö representing the period 6000-4500 yr B.P., indicating that an extensive pine forest grew in the area at that time. A striking feature in the Swedish Lapland graph is the gap from 3000 to 1500 yr B.P. Together with the valleys in the South Norwegian and Inari-Utsjoki curves it may be connected with the temperature drop discussed earlier.

The dates from Inari-Utsjoki form an undulating curve, which at first glance would seem to indicate an up-and-down variation in the pine limit. The vegetational zones in the area form a varying mosaic-type pattern at the present time, however, which also shows minor fluctuations in the level of the pine limit. The results are thus not an absolute indicator of fluctuations in the pine limit, but still attention should be paid to the marked drop at around 2500-2000 yr B.P., mentioned above.

5. Discussion and conclusions

Pine megafossils can be used as palaeoclimatological indicators provided that certain limitations are taken into account. Sufficient numbers of such fossils have been collected and dated in Fennoscandia to provide material for conclusions regarding climatic changes, and there are considerable differences between the various parts of Fennoscandia in terms of the main outlines of the development of their pine forests. The pine limit reached its highest level in the south much earlier than in the north, mainly on account of the spread of pine from the south during the phases following deglaciation. It is nevertheless possible that climatically the most favourable point of time for the growth of pines on the Scandinavian peninsula was as early as 9000-8000 yr B.P., i.e. before the species had even spread to Northern Fennoscandia. This may be explained by the relatively high amount of incoming solar radiation received by the northern areas in summertime, caused by astronomic factors, which became weaker towards the later Holocene. The high summer temperatures favourable for the growth of pines were a consequence of such a distribution of annual radiation.

The decline in the pine limit in the mountains of Southern Norway in 7000-5000 yr B.P. and the simultaneous maximum spread of pine in the north point to differences in climatic development between these regions. Temperatures possibly began to drop by around 7000 yr B.P. in Southern Norway, approx. 6000 yr B.P. in the Southern Swedish Scandes and probably some 2000 years later in the north. It is impossible to draw any final conclusions from the differences on the basis of the pine material available, but dendrochronological investigations will undoubtedly provide more information on this with reference to Northern Fennoscandia.

References

AAS, B. & FAARLUND, T. (1988): Postglasiale skog-grenser i sentrale sørnorske fjelltrakter. ^{14}C datering av subfossile furu og bjørkerester. (Summary: Postglacial forest limits in Central South Norwegian mountains. Radiocarbon datings of subfossil pine and birch specimens). Norsk Geogr. Tidsskr. 42, 25-61

BARTHOLIN, T. (1987): Dendrochronology in Sweden. Ann. Acad. Sci. Fenn. A III 145, 79-88

COHMAP members (1988): Climatic changes of the last 18,000 years: observations and model simulations. Science 241, 1043-1052

EKMAN, M. (1987): Postglacial uplift of the crust in Fennoscandia and some related phenomena. Int. Association of Geodesy, Section 5: Geodynamics. General Assembly in Vancouver, Canada. Stencil, 23 p.

ERONEN, M. (1979): The retreat of pine forest in Finnish Lapland since the Holocene climatic optimum: a general discussion with radiocarbon evidence from subfossil pines. Fennia 157/2, 93-114

ERONEN, M. (1988): A scrutiny of the Late Quaternary history of the Baltic Sea. Geol. Surv. Finl. Special Paper 6, 11-18

ERONEN, M. & HUTTUNEN, P. (1987): Radiocarbon-dated subfossil pines from Finnish Lapland. Geogr. Ann. 69A/2, 297-304

ERONEN, M.; HUTTUNEN, P. & ZETTERBERG, P. (1991): Opportunities for dendroclimatological research in Fennoscandia. Paläoklimaforschung 6, 81-92

ERONEN, M. & HYVÄRINEN, H. (1982): Subfossil pine dates and pollen diagrams from Northern Fennoscandia. Geol. Fören. Stockh. Förh. 103, 437-445

GEAR, A. J. & HUNTLEY, B. (1991): Rapid changes in the range limits of Scotch pine 4000 years ago. Science 251, 544-547

GROVE, J. M. (1988): The Little Ice Age. Methuen, London, 498 p.

HAFSTEN, U. (1981): An 8000 year old pine trunk from Dovre, South Norway. Norsk Geogr. Tidsskr. 35, 161-165

HUSTICH, I. (1948): The Scotch pine in northernmost Finland and its dependence on the climate in the last decades. Acta Bot. Fenn. 42, 1-75

HUSTICH, I. (1958): On the recent expansion of the Scotch pine in northernmost Europe. Fennia 82/3, 1-25

HYVÄRINEN, H. (1975): Absolute and relative pollen diagrams from northernmost Fennoscandia. Fennia 142, 1-23

HYVÄRINEN, H. (1976): Flandrian pollen deposition rates and tree line history in Northern Fennoscandia. Boreas 5, 163-175

HYVÄRINEN, H. (1993): Holocene pine and birch limits near Kilpisjärvi, Western Finnish Lapland: pollen stratigraphical evidence. (this volume)

IMBRIE, J. & IMBRIE, K. P. (1979): Ice Ages - solving the mystery. Macmillan, London, 224 p.

KARLÉN, W. (1976): Lacustrine sediments and tree limit variations as indicators of Holocene climatic fluctuations in Lapland: Northern Sweden. Geogr. Ann. 58A, 1-34

KULLMAN, L. (1987): Sequences of Holocene forest history in the Scandes, inferred from megafossil *Pinus sylvestris*. Boreas 16, 21-26

KULLMAN, L. (1988): Holocene history of the forest-alpine tundra ecotone in the Scandes Mountains (Central Sweden). New Phytol. 108, 101-110

KULLMAN, L. (1990a): Tree limit history during the Holocene in the Scandes Mountains, Sweden, inferred from subfossil wood. Rev. Palaeobot. and Palyn. 58, 163-171

KULLMAN, L. (1990b): Dynamics of altitudinal tree limits in Sweden: a review. Norsk Geogr. Tidsskr. 44, 103-116

KULLMAN, L. & ENGELMARK, O. (1990): A high Late Holocene tree limit and the establishment of the spruce forest-limit - a case study in Northern Sweden. Boreas 19, 323-331

KULLMAN, L. & HÖGBERG, N. (1989): Rapid natural decline of upper montane forests in the Swedish Scandes. Arctic 42, 217-226

LAMB, H. H. (1982): Climate, history and the modern world. Methuen, London, 387 p.

LUNDQVIST, G. (1959): ^{14}C-daterade tallstubbar från fjällen. Sver. Geol. Unders. C565, 1-21

LUNDQVIST, J. (1986): Late Weichselian glaciation and deglaciation in Scandinavia, Quat. Sci. Rev. 5, 269-292

MANGERUD, J. (1990): Paleoklimatologi. In: Drivhuseffekten og klimautviklingen. Norsk institutt for luftforskning. Rapport 21/90, 102-151

MOE, D. (1979): Tregrense-fluktuationer på Hardangervidda efter siste istid (Abstract: Tree line fluctuations on Hardangervidda during the last 9000 years). In: Fortiden i søkelyset. Datering med ^{14}C gjennom 25 år. NTH, Trondheim, 199-208

NESJE, A. & DAHL, S. O. (1991): Holocene glacier variations of Blåisen, Hardangerjökulen, Central Southern Norway. Quat. Res. 35, 25-40

NESJE, A.; KVAMME, M.; RYE, N. & LØVLIE, R. (1991): Holocene glacial and climate history of the Jostedalsbreen region, Western Norway: Evidence from lake sediments and terrestrial deposits. Quat. Sci. Rev. 10, 87-114

NUMMINEN, E. (1989): Metsäpuiden siemensato ja tuleentuminen Pohjois-Suomessa (Abstract: Seed crops of forest trees and ripening of seeds in North Finland). In: O. Saastamoinen & M. Varmola (eds.): Lapin Metsäkirja. Acta Lapp. Fenn. 15, 87-94

SELSING, L. & WISHMAN, E. (1984): Mean summer temperatures and circulation in a Southwest Norwegian mountain area during the Atlantic period, based upon changes of the alpine pine forest limit. Ann. Glaciol. 5, 127-132

Addresses of the authors:

Dr. M. Eronen, Department of Geology, University of Oulu, Linnanmaa, 90570 Oulu, Finland
Dr. P. Huttunen, University of Joensuu, Karelian Institute, P.O. Box 111, SF-80101 Joensuu, Finland

Dynamism of the altitudinal margin of the boreal forest in Sweden

Leif Kullman

Summary

On all studied time scales Swedish altitudinal tree limits have responded sensitively to climatic variability. Since the mid-Holocene period the main trend of pine tree limit change has been retreat and the tree limit ecotone has changed its character from predominance of evergreen pine to broad-leaved deciduous birch. Hypothetically, this shift basically reflects changed climatic seasonality in consequence of orbital variation. On century-scales and shorter, transient tree limit rises and recessions have been superimposed on the long-term trend of retreat. An episode of warming during the first half of the twentieth century induced a tree limit rise of about 40 m for *Betula pubescens* ssp. *tortuosa*, *Pinus sylvestris*, *Picea abies*, and *Sorbus aucuparia*. As a result of climatic cooling during the past 50 years, tree limits and the upper boreal forests have suffered severe climatic stress and dieback in some areas. These events could be studied in greater detail within a unique regional network of sites in the Southern Swedish Scandes, where baseline tree limit data were already collected in 1915/16 and a resurvey was carried out in the 1970s.

Zusammenfassung

In Schweden reagierten die höhenabhängigen Baumgrenzen zu allen untersuchten Zeitscheiben sehr empfindlich auf Klimaveränderungen. Seit dem mittleren Holozän zeigte die Baumgrenze einen rückläufigen Trend und die Zusammensetzung der baumgrenznahen Vegetation wandelte sich von einer Dominanz der immergrünen Kiefer zu der einer breitblättrigen, laubabwerfenden Birke. Vermutlich geht diese Veränderung im Grunde auf Änderungen der Erdbahnelemente zurück. Über die Jahrhunderte oder auch kürzere Zeiträumen hinweg gesehen, ist der langfristige Rückzugstrend der Baumgrenze zeitweilig durch vorübergehende Vorstöße überlagert worden. Eine Erwärmungsphase in der ersten Hälfte des zwanzigsten Jahrhunderts brachte für *Betula pubescens* ssp. *tortuosa*, *Pinus sylvestris*, *Picea abies* und *Sorbus aucuparia* einen Anstieg der Baumgrenze um etwa 40 m mit sich. Aufgrund einer Abkühlung während der letzten 50 Jahre gerieten die Baumgrenze und die nordborealen Wälder regional unter erheblichen ökologischen Druck und wurden vom Absterben bedroht. Diese Ereignisse konnten mit Hilfe eines einmaligen Netzwerkes von Fundpunkten in den südschwedischen Skanden sehr genau untersucht werden, wo bereits 1915/16 grundlegende Daten über Baumgrenzen gesammelt worden waren und wo in den 70-er Jahren eine erneute Untersuchung stattfand.

1. Introduction

The altitudinal zonation of the upper boreal forest in Sweden (and Fennoscandia) is characterized by a subalpine belt of mountain birch (*Betula pubescens* Ehrh. ssp. *tortuosa* (Ledeb.) Nyman), which is replaced at lower altitudes by montane coniferous forest with spruce (*Picea abies* (L.) Karst.) and pine (*Pinus sylvestris* L.). The tree limit of birch is about 50 to 100 m higher than that of spruce and pine respectively (KULLMAN, 1990). Tree species of secondary importance are *Sorbus aucuparia* L., *Alnus incana* (L.) Moench, *Populus tremula* L., *Prunus padus* L., and *Salix coaetanea* (Hartm.) Flod. Only *Sorbus aucuparia* reaches the same elevation as *Betula pubescens*. The vertical extent of the birch belt increases from the east towards the west. This is in close accord with the continentality-oceanicity gradient, implying increased depth and duration of the snow cover (HÄMET-AHTI, 1963, 1987; KULLMAN, 1981a).

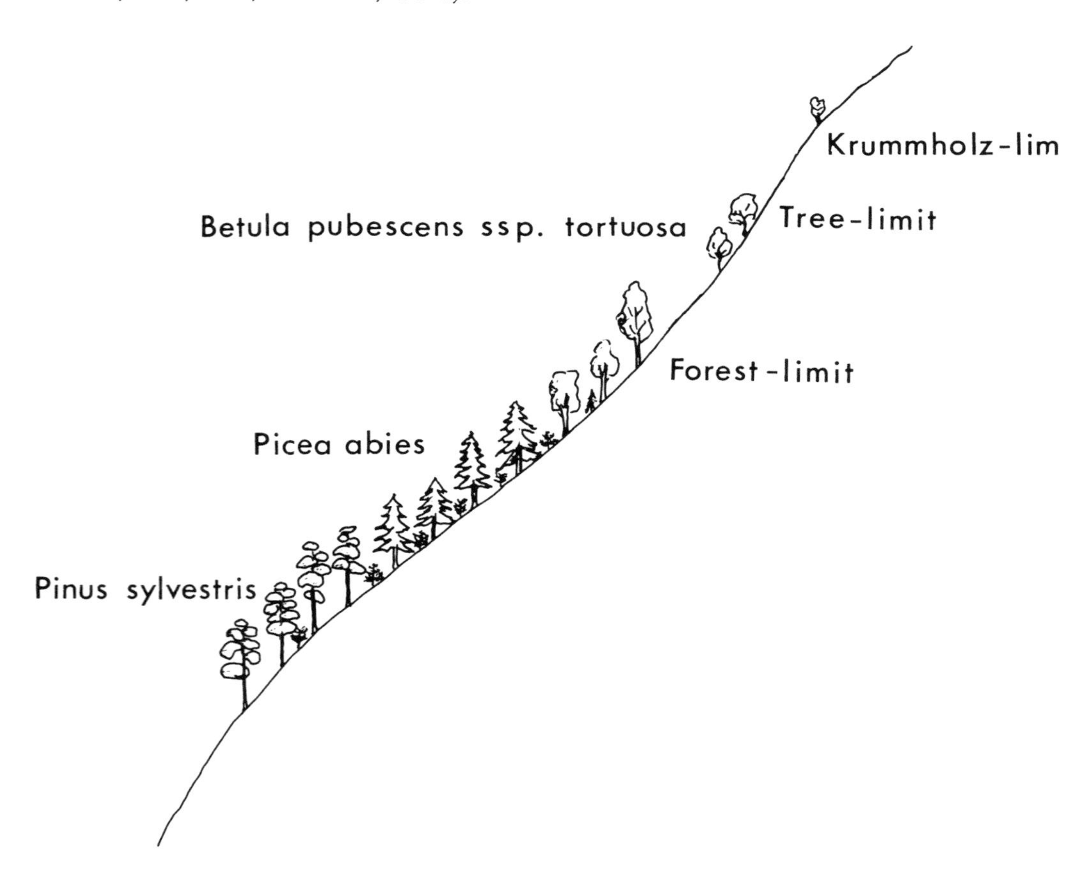

Fig. 1 Schematic pattern of altitudinal tree species zonation in the Swedish Scandes

Typically, the transition from closed forest to the barren alpine belt is gradual. This tension zone (the tree limit ecotone) is a kind of continuum of great ecological complexity and geographical variation as regards structure and function in consequence of both contemporaneous and historical ecological conditions (cp. HOLTMEIER, 1985). However, in order to facilitate comparisons both in space and time the tree limit may be defined strictly as the altitude of the uppermost individual with a minimum height of 2 m (cp. KULLMAN, 1990). Altitudinal dynamics of this fictive line, which runs evenly on most mountain slopes, is assumed to indicate the general trends of demographic population processes in this zone.

The current zonation pattern has developed successively during the Holocene in accordance with simulated changes of climatic seasonality and atmospheric circulation patterns (cp. KULLMAN, 1981a, 1989a; KUTZBACH & GUETTER, 1986; HUNTLEY, 1990).

The North Atlantic region appears to have been climatically very dynamic throughout the Holocene period (SHACKLETON et al., 1988; BRIFFA et al., 1990), although unambiguous support for coherent changes on the secular time scale is lacking (cp. WILLIAMS & WIGLEY, 1983). In response to this climatic dynamism, the tree limit ecotone has fluctuated widely with respect to structure, prevailing tree species, and elevation (HUSTICH, 1983; KULLMAN, 1990). The natural instability and the primaeval conditions make Sweden and Fennoscandia in general a suitable region for monitoring tree limits and thus learning about the fundamental relationships between climate and tree limits. Studies of recent and subrecent processes on population and ecosystem levels are prerequisites for using tree limit oscillations as palaeoclimatic proxy data.

This paper reviews the Swedish tree limit history during the Holocene. Emphasis is placed on the desirability of active research within an existing regional tree limit monitoring network founded on sites with old baseline data.

2. Pre-twentieth century tree limit history

Evidence from systematic surveys and radiocarbon dating of subfossil wood indicates that approx. 9000-7000 yr B.P. mainly pine made up the tree limit ecotone in the Southern Swedish Scandes (KARLÉN, 1976, 1988; KULLMAN 1987a,b, 1988a, 1989a, unpublished data). Pine then grew about 300 m higher altitudinally than it does at present (1975-90). The first signs of a distinct altitudinal belt of deciduous trees above the pine appeared around 7000 yr B.P., although small stands of both birch and alder had existed previously. Obviously, alder was a prominent feature in this belt, until approx. 5000 yr B.P. when birch gained increased dominance. Shortly after 6000 yr B.P. pine experienced an optimum period with maximum population density (Fig. 2). Birch extended its vertical distribution at least 200 m higher than currently. After about 5000 yr B.P., pine was successively replaced downslope by mountain birch. By analogy with present-day ecologies of pine, birch (Photo

1), and alder, the emergence of a subalpine birch forest belt was conceivably related to reduced climatic seasonality, implying increased ecological importance of abundant snow cover (Photo 2 and 3). Apparently, a birch belt evolved earlier in more western and northern parts of Fennoscandia (HYVÄRINEN, 1976; AAS & FAARLUND, 1988).

Recession of the pine tree limit has continued up to the present. Possibly, this apparently smooth process was punctuated by shorter episodes of both advance and accelerated retreat. Reasonably, such century-scale events are underrated. For some of the tree species these altitudinal shifts are merely qualitative manifestations, not necessarily involving a large number of trees in the critical zone (cp. KULLMAN, 1981b, 1990), and thus leaving a rather scanty record of pollen and megafossils. A few examples of short-term responses are provided below.

There are indications (subfossil wood) from northernmost Sweden of a marked but transient tree limit rise of pine during part of the period 300-600 A.D. (KARLÉN, 1988; KULLMAN & ENGELMARK, 1990). Warming during this period is corroborated by independent proxy data (LAMB, 1977).

The so-called Little Ice Age, from about the late sixteenth to the late nineteenth century, was the latest of several thermal declines on a secular scale during the Holocene period (GROVE, 1988, WIGLEY & KELLY, 1990). In Northern Fennoscandia the effects included recession of pine and birch tree limits, strongly reduced regeneration rates, and stand-level dieback of marginal boreal forest and, as consequence, the local formation of subalpine heaths (ERONEN, 1979; CASELDINE & MATTHEWS, 1987; KULLMAN 1987a, b, 1989a; JOSEFSSON, 1990). A similar episode, although perhaps shorter, with regional tree limit recession and increased geomorphic instability might have occurred around 6300 yr B.P. (KULLMAN, 1989a).

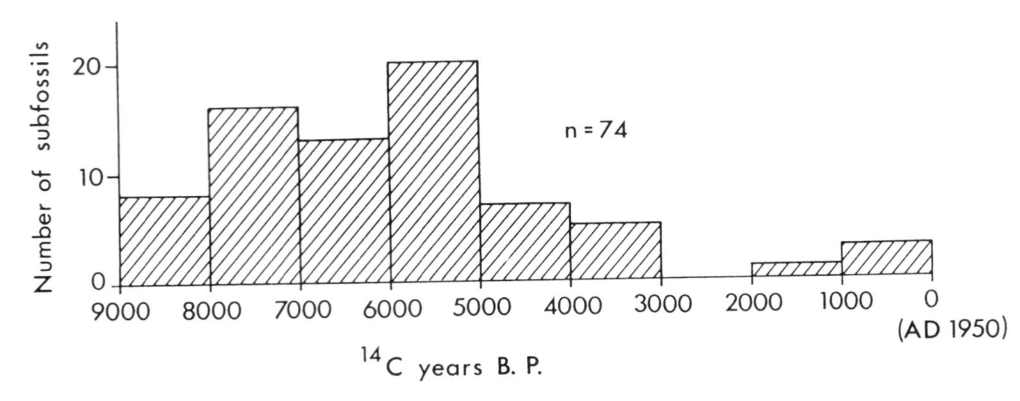

Fig. 2 Frequency/age distribution (^{14}C yr B.P.) of dates of megafossil pine above the present-day tree limit in the Southern Swedish Scandes. Data source: KULLMAN (1988a, 1989a)

Spruce is a late immigrant from the east and in the Scandes it has reached the present-day distributional limit during the past 2000 years or so in conformity with increased humidity (HUNTLEY & BIRKS, 1983; HAFSTEN, 1987; HUNTLEY, 1990). Its margin in Northern Sweden has obviously been in a dynamic equilibrium with prevailing climate for some recent centuries at least. Altitudinal advance took place during some periods of the nineteenth century, when temperature and soil moisture seem to have been optimally balanced (KULLMAN & ENGELMARK, 1990, 1991).

3. Marginal forests and climate during the twentieth century

3.1 Climatic variability

In Scandinavia and adjacent regions of the North the twentieth century has shown a great range of thermal variability, viz. warming until the 1940s and subsequent cooling approaching conditions at the end of the Little Ice Age (JONES et al., 1987; ALEXANDERSSON & ERIKSSON, 1989; VEDIN, 1990). The general trends for the air temperature in Northern Sweden (1860-1987) are displayed in Fig. 3.

3.2 Warming and tree limit rise

Responses of tree limits to present century climatic variability have been studied on a regional basis in the Southern Swedish Scandes (Fig. 4). Each assessment of the tree limit refers to a fixed transect. Hundreds of such determinations of the birch tree limit at regular intervals were conducted by SMITH (1920). During the early and mid-1970s these sites were resurveyed and a measure of altitudinal tree limit change (1915-1975) could be obtained (KULLMAN, 1979). The magnitude and the timing of tree limit change was further elucidated by age studies of the tree limit populations. Additionally, on many of these sites the recent tree limit history of *Pinus sylvestris*, *Picea abies*, and *Sorbus aucuparia* was outlined by age structure analyses (KULLMAN, 1979, 1990).

An average altitudinal tree limit rise of about 40 m for all studied tree species coincides with the post-Little Ice Age summer warming of about 1°C. The emergence of new tree-sized birches above the old tree limit correlates significantly with the mean temperature of June-August (Fig. 5). Radial growth of birch in the tree limit zone increased in congruence with summer warming (TRETER, 1984; KULLMAN, 1991c). Only for birch was the tree limit rise associated with abundance of new trees above the old tree limit. For other species, particularly pine, upslope displacement of the tree limit was accomplished by widely scattered individuals (KULLMAN, 1981b).

The frequency of sites with tree limit advance ranges between 50% and 75%, being highest for species capable of vegetative regeneration and long-term survival as krummholz and least for pine, with comparatively small ability in that respect. Apparently, tree limit rise of birch, spruce, and rowan was to some extent the result of phenotypic plasticity, i.e. progressive growth-form change of stunted krummholz (cp. KULLMAN, 1990). In some areas, however, spruce and pine have invaded by seeding and changed the character of previously birch dominated stands. This process was locally facilitated by disturbance from defoliating insects (KULLMAN, 1991a).

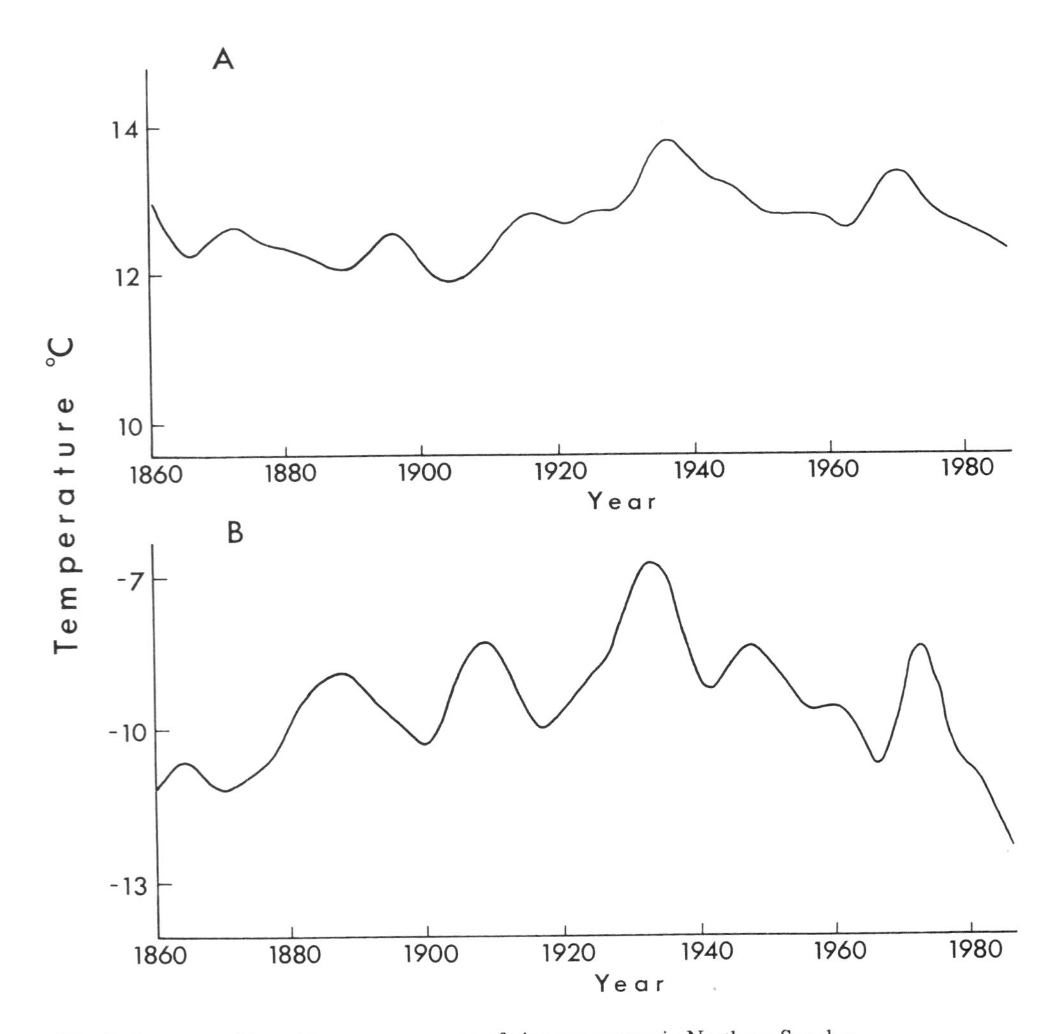

Fig. 3 Low-pass filtered ten-year-averages of air temperature in Northern Sweden. A) June-August. B) December-February. Source: ALEXANDERSSON & ERIKSSON (1989)

Fig. 4 Area in the Southern Swedish Scandes with a network of sites where the birch tree limit has been repeatedly monitored since 1915. Symbols: • sites where the tree limit advanced 1915-1975; * sites with stable tree limit. Source: KULLMAN (1979)

The variation in altitudinal response between nearby sites could be quite large, suggesting that the control of the tree limit elevation is not directly related everywhere to ambient air temperature. Evidently tree limits at certain sites are more responsive to climatic variability than elsewhere. It was particularly evident that the tree limits did not respond perceivably on the sites most exposed to wind and with sparse snow cover (KULLMAN, 1979, 1990).

The magnitude of birch tree limit rise was inversely proportional to the altitude of the early twentieth century tree limit altitude (KULLMAN, 1979). This suggests that some altitude dependent factor other than ambient air temperature, e.g. wind, may restrict tree-growth at high altitudes (cp. GRACE, 1989).

3.3 Recent cooling and tree limit stress

Case studies indicate that, in some areas, trees which contributed to the general tree limit advance earlier this century are suffering severe climatic stress, including crown dieback, stem breakage, and even individual mortality (KULLMAN, 1990). This process, which has been visible since the mid-1970s, is particularly pronounced for birch and spruce (Photo 4). Typically, individuals with almost total canopy defoliation still have vigorously growing basal foliage (spruce) or sprouts (birch and rowan). Possibly, such specimens (Photo 5) may survive in this state for many years, prepared to respond progressively to future warming episodes as in the first half of the twentieth century.

Pine growing on mineral soils has become more indifferent to the cooling trend and locally even advanced its tree limit by means of unchecked height growth of specimens established during the 1940s and 1950s (KULLMAN, 1988b). For all the tree species, seed regeneration has diminished substantially in the tree limit ecotone and mortality of smaller saplings has been high (KULLMAN, 1983, 1987b).

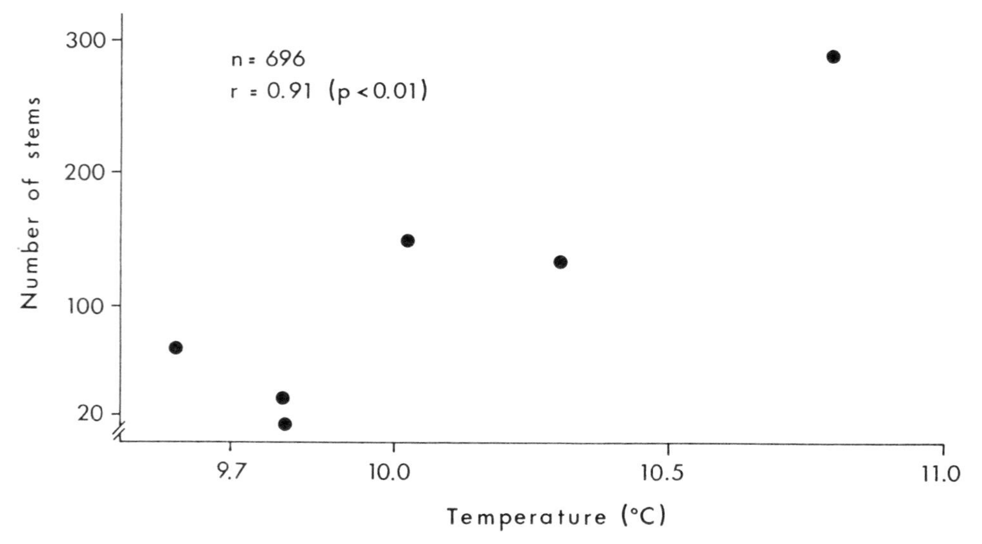

Fig. 5 Relationship between the numbers of new tree-sized stems of birch above the old (1915) tree limit (data from KULLMAN, 1979) and the June-August mean temperature. 10-year-age-classes of trees vs. decadal means of temperature, 1901-60. Temperature data are from the Storlien/Visjövalen meteorological station in the study area

Locally, after the mid-1980s, the dieback process has spread downslope to the upper closed coniferous forests. Stands of mature pine and spruce have suffered severe defoliation (reddish discoloration) particularly of unshaded, south-facing foliage above the maximum snow-limit (KULLMAN & HÖGBERG, 1989; KULLMAN, 1989b, 1991b), i.e. the typical symptoms of frost desiccation (cp. TRANQUILLINI, 1979; LARCHER & HÄCKEL, 1985).

These recessions are also manifested by decreasing tree-ring widths (TRETER, 1984; KULLMAN, 1989b, 1990, 1991b,c). In some populations the magnitude of defoliation is now so great that the radial growth will possibly be depressed for years or decades to come (cp. HUSTICH, 1978; SALEMAA & JUKOLA-SULONEN, 1990).

An important feature is the pronounced cooling of the early winter (December-February) mean temperature (Fig. 3). This is a very critical period of the year, when low air temperature in combination with a shallow snow cover may promote deep ground frost penetration and stress the ecosystem profoundly (cp. PRUITT, 1978). In fact, the recent dieback event coincides with episodic permafrost expansion and exceptionally late seasonal thawing of the ground, which related to extreme cold and sparse snow cover at the turn of 1986/87 in particular (KULLMAN, 1989c; JOSEFSSON, 1990). Besides, dieback was most spectacular on shallow peat soils and mineral soils with a thick mor humus layer, i.e. a setting ideal for ground frost preservation (KULLMAN, 1989b,c, 1991b). Obviously, winter conditions may be decisive for tree-growth at the altitudinal tree limits (cp. GRACE & NORTON, 1990; KULLMAN, 1990).

The recent regression of the margin of the boreal forest, which may be transient and spatially heterogeneous, fits into a more extensive pattern of intensified periglacial processes at the arctic and subarctic fringes of the North Atlantic region (e.g. KARTE, 1983; SCHUNKE, 1983; SVENSSON, 1986; RAPP & NYBERG, 1988; JOSEFSSON, 1990; NYBERG & LINDH, 1990). Reasonably, both biological and geomorphic systems in the periglacial environment are adjusting to a cooler climate, which could challenge ecological generalizations developed during warmer periods (KULLMAN & HÖGBERG, 1989; NYBERG & LINDH, 1990). An aspect of general relevance is that dramatic cooling of a few decades or less may reduce population density in the tree limit ecotone, lower tree limits and disintegrate biological structure developed during a much longer period of time. Thus, it may be hypothesized that the Holocene tree limit history and ecological succession have been much governed by short-term climatic extremes (cooling). From a uniformitarian point of view one should definitely anticipate more oscillating tree limits than what is evident from available palaeoecological records. Recent developments emphatically demonstrate that mechanisms for such short-term instability exist. Speculations of vegetation responses to future climatic change must take the profoundly stressed state of northern and high-elevation ecosystems at the present into consideration, which will delay responses to possible future warming and lower the physiological tolerance of continued cooling.

Photo 1 Pine preferentially grows on relatively dry sites with a sparse snow cover. Birch finds optimal conditions where the snow is deeper and more persistent. Reasonably, the historical shifts in relative abundances of these species reflect climatic oscillations related to the above variables

Photo 2 Prior to approx. 6000 yr B.P. pine grew at this site, above the current tree limit where deep and practically perennial snow fields prevail today. Mt. Helagsskaftet, 945 m a.s.l., 3 September 1986

Photo 3 Remnant of a pine trunk which was found just outside the snow border on photo 2. Radiocarbon dating yielded 6950±95 yr B.P. (ST-10833)

Photo 4 Sparse spruce population close to the tree limit on Mt. Kuormakka (430 m a.s.l.) in northernmost Sweden. During the 1980s the trees have simultaneously suffered severe defoliation. 17 September 1987

Photo 5 Tree-limit birch whose main stem died in recent years

4. Future monitoring of tree limits in the Swedish Scandes

This review provides evidence that Swedish tree limits trace climatic warming and cooling sensitively, even on a decadal time-scale. In order to develop transfer functions between climatic variability and the tree limit behaviour, future tree limit dynamics should be monitored within the region of the Southern Swedish Scandes, where extensive baseline data have been collected since the early twentieth century. Besides, pertinent information concerning the long-term Holocene tree limit history, general vegetation patterns, geology, climate, etc. are available here. This is, in fact, a unique starting point for monitoring, without counterpart in any part of the world (cp. KULLMAN, 1990).

References

AAS, B. & FAARLUND, T. (1988): Postglaciale skoggrenser i sentrale sørnorske fjelltrakter. Norsk Geogr. Tidsskr. 42, 25-61

ALEXANDERSSON, H. & ERIKSSON, B. (1989): Climate fluctuations in Sweden 1860-1987. SMHI Reports Meteorology and Climatology 58, 1-54

BRIFFA, K. R.; BARTHOLIN, T. S.; ECKSTEIN, D.; JONES, P. D.; KARLÉN, W.; SCHWEINGRUBER, F. H. & ZETTERBERG, P. (1990): A 1,400-year tree-ring record of summer temperatures in Fennoscandia. Nature 346, 434-439

CASELDINE, C. J. & MATTHEWS, J. A. (1987): Podzol development, vegetation change and glacier variations at Haugabreen, Southern Norway. Boreas 16, 215-230

ERONEN, M. (1979): The retreat of pine forest in Finnish Lapland since the Holocene climatic optimum: a general discussion with radiocarbon evidence from subfossil pines. Fennia 157, 93-114

GRACE, J. (1989): Tree lines. Phil. Trans. Roy. Soc. London B 324, 233-245

GRACE, J. & NORTON, D. A. (1990): Climate and growth of *Pinus sylvestris* at its upper altitudinal limit in Scotland: evidence from tree growth-rings. J. Ecol. 78, 601- 610

GROVE, J. M. (1988): The Little Ice Age. Methuen, New York, 498 p.

HAFSTEN, U. (1987): Vegetasjon, klima og landskaps-utvikling i Trøndelag etter siste istid. Norsk Geogr. Tidsskr. 41, 101-120

HÄMET-AHTI, L. (1963): Zonation of the mountain birch forests in northernmost Fennoscandia. Ann. Bot. Soc. Zool. Bot. Fenn. 'Vanamo' 34, 1-127

HÄMET-AHTI, L. (1987): Mountain birch and mountain birch woodland in NW Europe. Phytocoenologia 15, 449-453

HOLTMEIER, F.-K. (1985): Die klimatische Waldgrenze - Linie oder Übergangssaum (Ökoton)? Erdkunde 39, 271-285

HUNTLEY, B. (1990): European post-glacial forests: compositional changes in response to climatic change. J. Veget. Sci. 1, 507-515

HUNTLEY, B. & BIRKS, H. J. B. (1983): An atlas of past and present pollen maps for Europe: 0-13000 years ago. Cambridge Univ. Press, Cambridge, 667 p.

HUSTICH, I. (1978): The growth of Scots pine in Northern Lapland, 1928-77. Ann. Bot. Fenn. 15, 241-259

HUSTICH, I. (1983): Tree-line and tree-growth studies during 50 years: some subjective observations. Nordicana 47, 241-259

HYVÄRINEN, H. (1976): Flandrian pollen deposition rates and tree-line history in Northern Fennoscandia. Boreas 5, 163-175

JONES, P. D.; WIGLEY, T. M. L.; FOLLAND, C. L. & PARKER, D. E. (1987): Spatial patterns in recent worldwide temperature trends. Climate Monitor 16, 175-180

JOSEFSSON, M. (1990): The geoecology of subalpine heaths in the Abisko Valley, Northern Sweden. UNGI Rapport 78, 1-180

KARLÉN, W. (1976): Lacustrine sediments and tree limit variations as indicators of Holocene climatic fluctuations in Lappland, Northern Sweden. Geogr. Ann. 58A, 1-34

KARLÉN, W. (1988): Scandinavian glacial and climatic fluctuations during the Holocene. Quat. Science Rev. 7, 199-209

KARTE, J. (1983): Periglacial phenomena and their significance as climatic and edaphic indicators. GeoJournal 7, 329-334

KULLMAN, L. (1979): Change and stability in the altitude of the birch tree limit in the Southern Swedish Scandes 1915-1975. Acta Phytogeogr. Suec. 65, 1-121

KULLMAN, L. (1981a): Some aspects of the ecology of the Scandinavian subalpine birch forest belt. Wahlenbergia 7, 99- 112

KULLMAN, L. (1981b): Recent tree limit dynamics of Scots pine (*Pinus sylvestris* L.) in the Southern Swedish Scandes. Wahlenbergia 8, 1-67

KULLMAN, L. (1983): Short-term population trends of isolated tree limit stands of *Pinus sylvestris* L. in central Sweden. Arct. Alp. Res. 15, 369-382

KULLMAN, L. (1987a): Little Ice Age decline of a cold marginal *Pinus sylvestris* forest in the Swedish Scandes. New Phytol. 106, 567-584

KULLMAN, L. (1987b): Long-term dynamics of high-altitude populations of *Pinus sylvestris* in the Swedish Scandes. J. Biogeogr. 14, 1-8

KULLMAN, L. (1988a): Holocene history of the forest-alpine tundra ecotone in the Scandes Mountains (central Sweden). New Phytol. 108, 101-110

KULLMAN, L. (1988b): Short-term dynamic approach to tree limit and thermal climate: evidence from *Pinus sylvestris* in the Swedish Scandes. Ann. Bot. Fenn. 25, 219-227

KULLMAN, L. (1989a): Tree-limit history during the Holocene in the Scandes Mountains, Sweden, inferred from subfossil wood. Rev. Palaeobot. Palyn. 58, 163-171

KULLMAN, L. (1989b): Cold-induced dieback of montane spruce forests in the Swedish Scandes - a modern analogue of paleoenvironmental processes. New Phytol. 113, 377-389

KULLMAN, L. (1989c): Geoecological aspects of episodic permafrost expansion in North Sweden. Geogr. Ann. 71A, 255-262

KULLMAN, L. (1990): Dynamics of altitudinal tree limits in Sweden: a review. Norsk Geogr. Tidsskr. 44, 103-116

KULLMAN, L. (1991a): Structural change in a subalpine birch woodland in North Sweden during the past century. J. Biogeogr. 18, 53-62

KULLMAN, L. (1991b): Cataclysmic response to recent cooling of a natural boreal *Pinus sylvestris* forest in Northern Sweden. New Phytol. 117, 351-360

KULLMAN, L. (1991c): Pattern and process of present tree limits in the Tärna region, Southern Swedish Lapland. Fennia 169(1), 25-38

KULLMAN, L. & ENGELMARK, O. (1990): A high late Holocene pine tree limit and the establishment of the spruce forest-limit - a case study in Northern Sweden. Boreas 19, 323-331

KULLMAN, L. & ENGELMARK, O. (1991): Historical biogeography of *Picea abies* at its subarctic limit in Northern Sweden. J. Biogeogr. 18, 63-70

KULLMAN, L. & HÖGBERG, N. (1989): Rapid natural decline of upper montane forests in the Swedish Scandes. Arctic 42, 217-226

KUTZBACH, J. E. & GUETTER, P. J. (1986): The influence of changing orbital parameters and surface boundary conditions on climate simulations for the past 18,000 years. J. Atmosph. Sci. 43, 1726-1759

LAMB, H. H. (1977): Climate: present, past and future. Vol. 2: Climatic history and the future. Methuen, London, 835 p.

LARCHER, W. & HÄCKEL, H. (1985): Handbuch der Pflanzenkrankheiten. Bd. 1, Paul Parey, Berlin, 326 p.

NYBERG, R. & LINDH, L. (1990): Geomorphic features as indicators of climatic fluctuations in a periglacial environment, Northern Sweden. Geogr. Ann. 72A, 203-210

PRUITT, W. O. Jr. (1978): Boreal ecology. Edward Arnold, London, 73 p.

RAPP, A. & NYBERG, R. (1988): Mass movements, nivation processes and climatic fluctuations in northern Scandinavian mountains. Norsk Geogr. Tidsskr. 42, 245-253

SALEMAA, H. & JUKOLA-SULONEN, E.-L. (1990): Vitality rating of *Picea abies* by defoliation class and other vigour indicators. Scandinavian J. For. Res. 5, 413-426

SCHUNKE, E. (1983): Aktuelle Palsabildung in der Subarktis und ihre klimatischen Bedingungen. Abh. Akad. Wiss. Göttingen, Mathemat.-Physikal. Klasse, 3. Folge 35, 19-33

SHACKLETON, N. J.; WEST, R. G. & BOWEN, D. Q. (eds.) (1988): The past three million years: evolution of climatic variability in the North Atlantic Region. Phil. Trans. Roy. Soc. London B 318, 409-688

SMITH, H. (1920): Vegetationen och dess utvecklingshistoria i det centralsvenska högfjällsområdet. Almqvist & Wiksells, Uppsala, 238 p.

SVENSSON, H. (1986): Permafrost. Some morphoclimatic aspects of periglacial features of northern Scandinavia. Geogr. Ann. 68A, 123-130

TRANQUILLINI, W. (1979): Physiological ecology of the alpine timberline. Springer, Berlin, 131 p.

TRETER, U. (1984): Die Baumgrenzen Skandinaviens. Franz Steiner, Wiesbaden, 111 p.

VEDIN, H. (1990): Frequency of rare weather events during periods of extreme climate. Geogr. Ann. 72A, 151-155

WIGLEY, T. M. L. & KELLY, P. M. (1990): Holocene climatic change, ^{14}C wiggles and variations in solar irradiance. Phil. Trans. Roy. Soc. London B 330, 547-558

WILLIAMS, L. D. & WIGLEY, T. M. L. (1983): A comparison for Late Holocene summer temperature variations in the northern hemisphere. Quat. Res. 20, 286-307

Address of the author:

Dr. L. Kullman, Department of Physical Geography, University of Umeå, S-901 87 Umeå, Sweden

Climate and growth of mountain birch near the treeline in Northern Sweden and Iceland

Bjartmar Sveinbjörnsson

Summary

Growth of mountain birch decreases with increasing elevation in Iceland and Swedish Lapland and this growth decrease can be reduced or eliminated by nitrogen fertilizer application. This nutrient limitation at higher elevations appears to be related to reduced soil nitrogen mineralization brought about by lower litter quality at higher elevations and perhaps a different decomposer flora. The reduced tree growth and nitrogen mineralization is not related to summer climatic conditions as temperatures are actually lowest at low elevations in mid summers and precipitation and soil moisture increase somewhat with increasing elevation. Climatic variables, not measured in these studies, such as winter conditions may be more important in dilimiting ground flora distribution along topographic gradients which in turn indirectly affect tree growth through litter production of varying quality resulting in varying mineralization.

Zusammenfassung

In Island und Schweden nimmt die Wuchshöhe der Moorbirke mit zunehmender Höhe über dem Meeresspiegel ab, wobei die Reduzierung des Höhenwachstums durch die Gabe von Stickstoffdüngern gemildert bzw. aufgehoben werden kann. Die Verminderung des Nährstoffangebotes in größerer Höhe hängt vermutlich mit einer gehemmten Mineralisierung des Bodenstickstoffs zusammen, was auf eine geringere Qualität der Streu und wahrscheinlich eine veränderte Zusammensetzung der Dekompositionsflora in größerer Höhe zurückgeht. Das eingeschränkte Baumwachstum und die verminderte Stickstoffmineralisierung sind nicht auf die sommerlichen Witterungsverhältnisse zurückzuführen, da die Temperaturen im Hochsommer in tieferen Lagen *de facto* am niedrigsten liegen und Niederschläge sowie Bodenfeuchtigkeit mit steigender Höhe leicht zunehmen. Andere, hier nicht berücksichtigte Klimafaktoren wie beispielsweise die winterlichen Verhältnisse wirken sich vermutlich durch die Verteilung und Limitierung des Bodenbewuchses entlang topographischer Gradienten aus. Der Bodenbewuchs beeinflußt das Baumwachstum indirekt, da die jeweils produzierte Streu von unterschiedlicher Qualität sein kann und damit auch ein unterschiedliches Potential für die Mineralisierungsprozesse bereitstellt.

1. Introduction

Climatic variables correlate with growth of tree species and these correlations are particularly strong near the treeline (FRITTS, 1976; JACOBY, 1983). The very basis for climate reconstruction from tree-rings on one hand and climatic control of the position of treelines on the other hand (FRITTS, 1976) is that the climate affects directly various physiological processes which control growth. Of these variables, temperature is of particular interest at northern treelines as severe water stress of mature trees is rare there both in summer (AUGER, 1974; VOWINCKEL et al., 1975; BLACK & BLISS, 1980; SVEINBJÖRNSSON, 1983, unpubl. data) and winter (author's unpublished data).

Summer temperatures of shoot meristems have been shown to decrease with elevation (GRACE et al., 1989) and have been assumed to play a role in reduced tree growth at treeline by regulating growth respiration (DAHL & MORK, 1959; SKRE, 1972). Conversely, low temperature has also been suggested to limit production of respiratory substrate through the reduction of the instantaneous photosynthetic rate (TRANQUILLINI, 1979) as has vapor pressure deficit (GOLDSTEIN, 1981). Data supporting these or other hypotheses concerning causal relationships are hard to find. Under equal conditions, mid-season rates of photosynthesis and dark respiration of larch (*Larix laricina*) and black spruce (*Picea mariana*) in Northern Quebec (AUGER, 1974; VOWINCKEL, 1975) and of white spruce saplings in Alaska (*Picea glauca*) (ABADIE, 1991) do not differ between treeline plants and those further below where growth is greater.

Several temperature averages such as July or summer mean, maximum, or minimum air temperatures have been related to treeline positions as have various heat accumulation indices (KÖPPEN, 1936). These may have relevance if climate turns cooler with increasing elevation. Even if that it not the case, the duration of the activity season is likely to be compressed. In relation to the photosynthetic season, earlier budburst at lower elevations can be observed for deciduous trees but less is known of evergreen trees (see though TRANQUILLINI, 1979; ABADIE, 1991). Treeline location has been shown to correlate well with climatically derived respiration equivalents in Norway (SKRE, 1972) but also with its functional opposite, climatically based calculation of assimilate production in Canada (VOWINCKEL, 1975).

Soil or root zone spring and early summer temperatures have been demonstrated to be a more precise predictor of bud flushing and shoot extension growth than air temperatures both at the polar treeline in Canada (SCOTT et al., 1987) and at subarctic and temperate altitudinal treelines (ABADIE, 1991; HANSEN-BRISTOW, 1986). The effect of low root zone temperature on plant growth and by extension tree growth at treeline has been attributed to reduced water uptake (DÖRING, 1935 as cited in LARCHER, 1973) both through increased viscosity of water and through reduced root water uptake. Lower photosynthetic rates have been found in trees growing near late lying snow patches in the Rocky Mountains (DELUCIA & SMITH, 1987). Similarly root nutrient uptake is temperature sensitive albeit less so in species growing in the subarctic zone (TRYON & CHAPIN, 1983). While total soil nitrogen content has both been shown to increase and decrease with altitude (EHRHARDT, 1961; MASUZAWA, 1985), soil nutrient availability may decrease with elevation through reduced mineralization irrespective of soil nutrient content.

The use of leaf or twig CO_2 exchange rates, or root water or nutrient uptake rates for explaining whole plant performance ignores possible shifts in area, volume or weight of different plant parts. It is well known that root/shoot ratios in communities change along gradients of environmental severity (BLISS, 1971) and KÖRNER & RENHARDT (1987) showed that perennial herbaceous plants from high altitude in the Alps when compared to low altitude plants allocated 15% less to above ground parts, increased specific root length by 50%, and tripled mean individual rooting density. However, the total plant uptake, use, or storage of carbon, water and nutrients is often difficult to assess, especially in field grown plants, let alone mature trees.

In addition to these problems, there are several other complications in attributing the effects of climate on tree growth and treeline position. One is the well documented carry-over effects from one year to another. These carry-over effects on growth need not be directly caused by climate, but may be caused by varying reproductive efforts, insect grazing, or pathogens (ROHMEDER, 1967).

2. Subarctic mountain birch studies

Below, I will summarize some of the findings of treeline related research on mountain birch (*Betula pubescens* Ehr. ssp. *tortuosa* (Ledeb.) Nyman) which I, my students and collaborators have studied for more than a decade in Swedish Lapland (SVEINBJÖRNSSON, 1983; SVEINBJÖRNSSON et al., 1992a; DAVIS et al., 1991) and for half a decade at two sites in Iceland (SVEINBJÖRNSSON et al., 1992b).

2.1 Growth

Height growth (Fig. 1; SVEINBJÖRNSSON et al., 1992a) decreases with elevation. When comparing trees in Swedish Lapland, those above the forest limit or forest border (505 to 630 m a.s.l. depending on location) always grow significantly less in height than those in the valley (360 to 385 m a.s.l.). In some years there is also a significant difference between trees above and at the forest border.

This altitudinal decrease in growth relates to a delayed bud burst at higher elevations (SVEINBJÖRNSSON, unpubl. data) but not necessarily lower air or soil temperatures during the growing season (Table 1, DAVIS et al., 1991). In fact, July air and soil temperatures are lowest in the lowest altitudinal zone. The time of budburst and subsequent conditions vary from year to year. Sometimes bud burst occurs early, i.e. at the end of May or in early June, only to be followed by colder temperatures or snowstorms. In the first instance the leaf expansion might be completed at the lower elevations but drag on at higher elevations, while in the latter altitudinally increasing leaf loss would take place. In a rapid and uninterrupted onset of warmer temperatures the altitudinal differences of growth and leaf expansion are minor. Fall colouration and leaf fall happens at about the same time at all elevations.

Table 1 Soil temperature and precipitation during midsummer and early fall of 1988 and 1989 in three different altitudinal zones on Mount Nuolja (Njulla), Swedish Lapland. (By permission from Arctic and Alpine Research)

Location	T max [a] (°C)	T min [b] (°C)	Precipitation (cm)
		1988 (2 July - 1 September)	
Valley	8.0 (11.1)	6.8 (4.2)	9.4
Forest limit	16.7 (22.0)	12.0 (9.0)	8.3
Treeline	22.2 (26.5)	7.9 (5.0)	8.0
		1989 (12 July - 1 September)	
Valley	9.0 (11.3)	7.6 (6.0)	8.5
Forest limit	15.1 (20.5)	5.9 (2.6)	11.1
Treeline	16.1 (19.7)	7.5 (2.6)	9.5

a Seasonal maximum in parentheses
b Seasonal minimum in parentheses

2.2 Carbon balance

The causes of these altitudinal growth decreases have been studied. Maximum instantaneous photosynthetic rates do not vary significantly between altitudinal zones averaging about 14 mg CO_2 g^{-1} h^{-1} for detached branches (SVEINBJÖRNSSON, 1983), and varying from 8 to 12 mg g^{-1} h^{-1} for attached branches under natural light and temperature conditions

(Fig. 2). Dark respiration rates of attached branches do not differ between the two upper elevations. When early-season snowstorms at the treeline occur and result in total leaf loss the calculated seasonal carbon balance there is reduced by about 30% (SVEINBJÖRNSSON, unpubl. data). However, carbohydrate concentrations do not vary significantly with elevation (SVEINBJÖRNSSON et al., 1992a) and thus there is no direct evidence for either photosynthetic or respiratory reduction or increase at higher elevations.

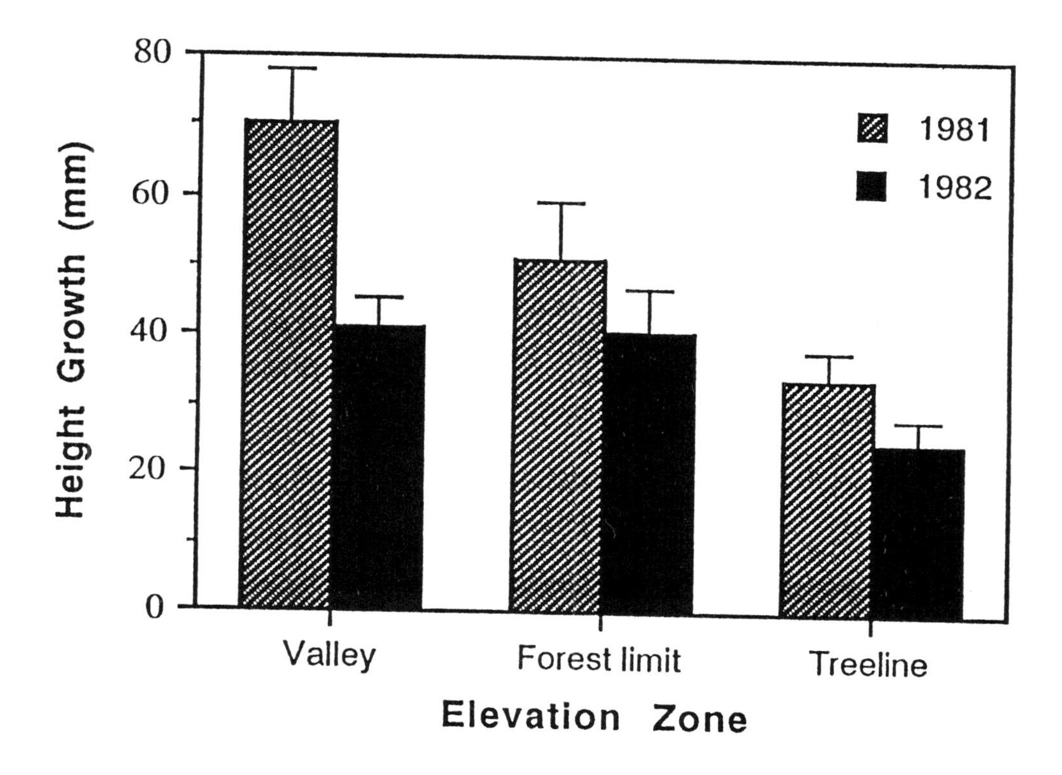

Fig. 1 Height growth of mountain birch in three altitudinal zones, valley, forest limit, and treeline on Mount Luovare, Swedish Lapland. The bars represent zone means and the lines standard error of zone means. (By permission from Blackwell Scientific)

2.3 Plant and soil nutrients

Little is known of the acquisition of nutrients along altitudinal gradients. While KARLSSON & NORDELL (1988a) found that low soil temperature significantly reduced mountain birch sapling uptake of nitrogen, foliar nitrogen concentrations increase with elevation in mountain birch in Swedish Lapland and in Iceland (SVEINBJÖRNSSON et al., 1992a,b) in concurrence with generally observed patterns (SVEINBJÖRNSSSON, 1992). The faster growing Icelandic birches have higher nitrogen concentrations than the ones in Swedish Lapland, so that geographically there is a positive relationship between leaf nitrogen and height growth, while along local altitudinal gradients, there is a negative relationship between leaf nitrogen and height growth (SVEINBJÖRNSSON et al., 1992b). KARLSSON & NORDELL (1988b) have found a positive relationship between maximum instantaneous net photosynthetic rate and foliar nitrogen concentration, with the highest rates in selected upper elevation birches. However, statistical analysis of the photosynthetic rates of in situ trees in different altitudinal zones, did not show significant altitudinal variation (Table 2, SVEINBJÖRNSSON, 1983).

Table 2 Net photosynthetic rates of excised branches of mountain birch from different zones on Mount Nuolja. (By permission from Centre d'études nordiques, Laval University, Québec)

Net photosynthesis (mg Co_2 g^{-1} h^{-1}) of excised branches of *Betula pubescens* (means ± standard deviation): A comparison between tree size and location[1]

	Tree-line	Forest-limit		Abisko Valley	
Tree size	Small	Small	Large	Small	Large
Mean	12.92±4.36	16.53±4.52	14.21±6.87	12.64±4.66	11.95±5.33
Number of samples	11	7	9	8	5

[1] The measurements were made at radiant density flux of 600 µE m^{-2} s^{-1} (400-700 nm) and 15°C ± 1.3°C

Soil nutrient availability may well decrease with elevation and hence limit tree growth. In a study of soil carbon and nitrogen turnover in Swedish Lapland we found (DAVIS et al., 1991) that there was a sharp decrease in mean soil nitrogen concentration (Table 3) and soil nitrogen mineralization with elevation. Neither soil temperature nor soil water content is lower at higher elevations but litter decomposability does vary. Thus, birch leaf litter from treeline decomposes slower than valley and forest limit birch leaf litter even when transplanted to lower elevations. In addition, the ground flora may change with elevation and contribute further to total system litter decomposability changes. Soil respiration rates also dropped markedly above the forest border and this may be caused by reduced decomposer numbers and/or diversity of decomposer taxa.

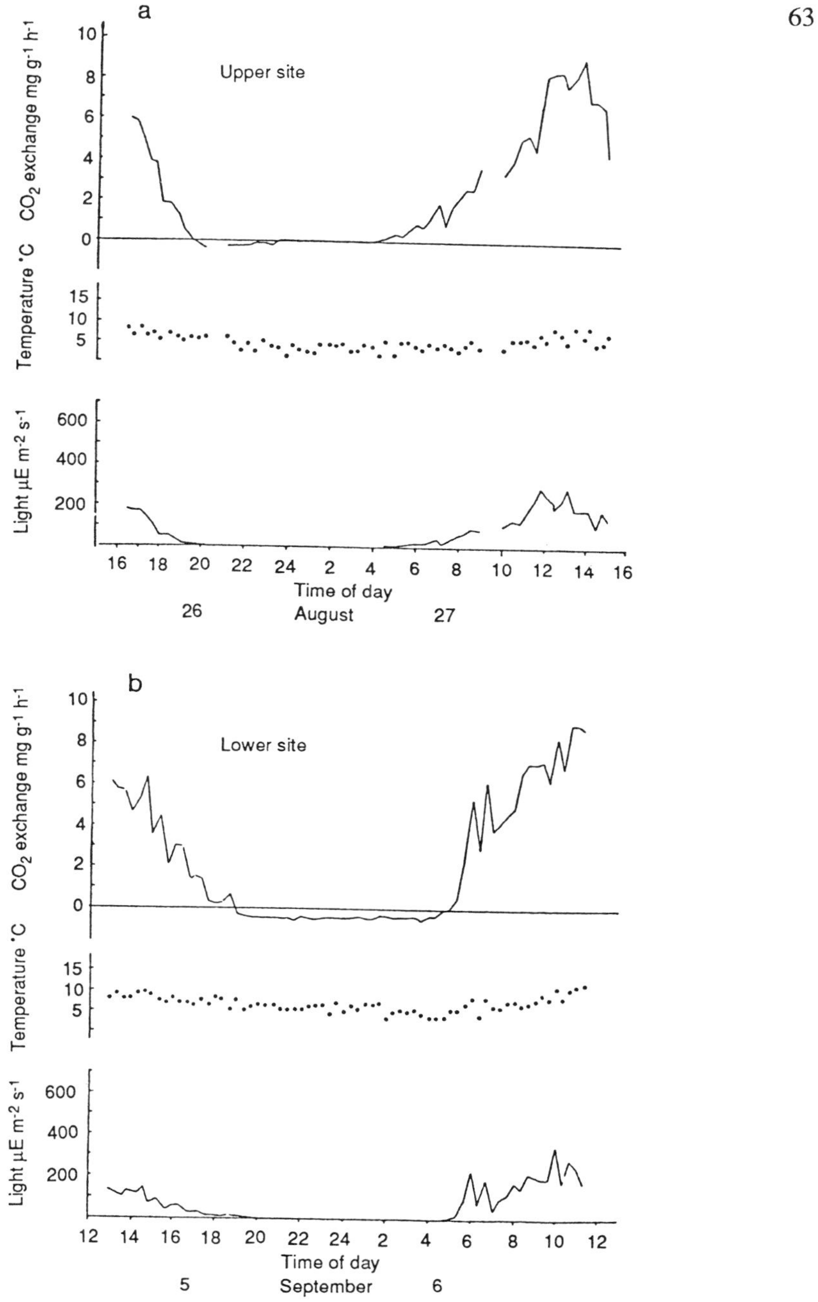

Fig. 2 Net instantenous photosynthetic rate, photon flux density and leaf temperature of mountain birch trees growing *in situ* at the forest limit and above it in the treeline zone on Mount Nuolja, Swedish Lapland. (By permission from Centre d'Études Nordiques, Laval University, Québec)

Table 3 Soil ammonium concentrations (means for July 1988 to June 1989 and seasonal range (means per zone)

Zone	Mean	Seasonal range
Valley	148 $\mu g\ g^{-1}$	87-218 $\mu g\ g^{-1}$
Forest limit	190 $\mu g\ g^{-1}$	109-328 $\mu g\ g^{-1}$
Treeline	11 $\mu g\ g^{-1}$	3-9 $\mu g\ g^{-1}$

If soil nutrient availability is increasingly limiting tree growth as one approaches treeline, then addition of soil nutrients should affect trees more at higher than lower elevations. An altitudinal fertilizer experiment in Swedish Lapland (Fig. 3, SVEINBJÖRNSSON, 1992) showed that this was the case for nitrogen application while phosphorus application did not consistently increase growth. Complete fertilizer stimulated growth in accordance with its nitrogen component dose. The effects of the fertilizer application lasted for several years. The mechanism for this growth increase has not been demonstrated but is likely to involve allocation shifts from root to shoot growth at higher elevations whereas leaf area increased in a similar manner at all altitudes.

Plant reproductive effort is also affected by soil nutrient availability. In Sweden, male catkin production was stimulated by phosphorus application while female inflorescence and hence potential seed set was not significantly affected by any nutrient treatment. In Iceland however, nitrogen application reduced the number of non-reproducing trees. Of importance for climate reconstruction there was no link between growth and female inflorescence number even when considering time lags.

2.4 Biotic and catastrophic factors

The autumn moth (*Oporinia autumnata* Bkh.) does cause altitudinal variation in mountain birch defoliation leading to reduced tree growth and in extreme cases to tree death (TENOW, 1975). The effect is however greatest at mid-altitudes as low winter temperatures at high and low elevations are lethal to overwintering moth eggs. New shoot damage and death due to fungal infections are common both in Sweden and Iceland but no altitudinal trends have been found (author's unpubl. data). Ptarmigan grazing damage to new shoots of young plants have been found at both high and low forest borders in Iceland. Sheep and reindeer grazing has been demonstrated to adversely affect birch forests (THORSTEINSSON, 1986; EMANUELSSON, 1987) and both have probably lowered treeline, the former in Iceland and the latter in Swedish Lapland. Catastrophic physical events such as snow and ice loads and tree breakage are common in the oceanic climate in Iceland while forest fires, important in North America, are rare and limited in extent in Scandinavia and absent in Iceland.

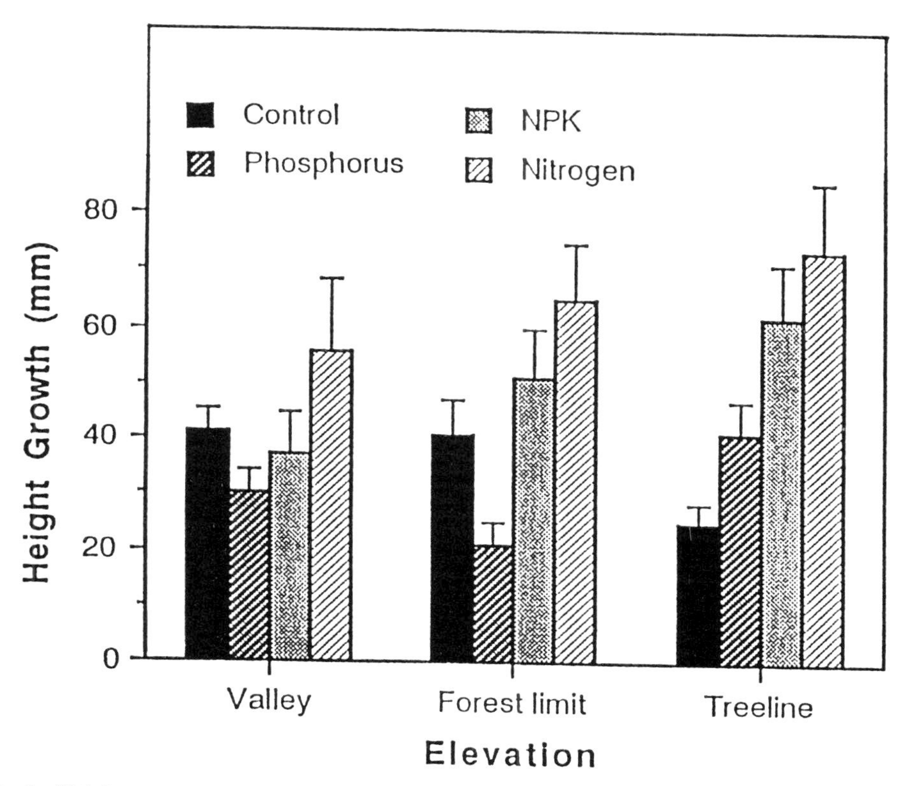

Fig. 3 Height growth of mountain birch trees in different elevational zones on Mount Luovare, Swedish Lapland and receiving different fertilizers. (By permission from Blackwell Scientific)

Acknowledgements

Financial support fom the U.S. National Science Foundation (BSR 8706510), the Swedish Academy of Sciences, the Swedish Natural Science Fund, and the Icelandic National Science Foundation is greatfully acknowledged. Logistic support was given by the Abisko Scientific Research Station and Skógræktarfélag Eyfirðinga.

References

ABADIE, W. D. (1991): Growth and carbon flux of white spruce at different elevations in the Chugach Mountains, Alaska. M. Sc. Thesis, Univ. of Alaska Anchorage, 97 p.

AUGER, S. (1974): Growth and photosynthesis of *Larix laricina* (Du Roi) D. Koch in the Subarctic at Schefferville, Quebec. M. Sc. Thesis, McGill Univ., Montréal, 114 p.

BLACK, R.A. & BLISS, L. C. (1980): Reproductive ecology of *Picea mariana* (Mill) B.S.P., at tree line near Innvik, Northwest Territories, Canada. Ecol. Monogr. 50, 331-354

BLISS, L. C. (1971): Arctic and alpine plant life cycles. Ann. Rev. Ecol. Syst. 2, 405-438

DAHL, E. & MORK, E. (1959): Om sambandet mellom temperatur, andning, og vekst hos gran (*Picea abies* (L.) Karst.). Medr. Nor. SkogforsVes. 16, 81-93

DAVIS, J.; SCHOBER, A.; BAIN, M. & SVEINBJÖRNSSON, B. (1991): Soil carbon and nitrogen turnover at and below the elevational treeline in Northern Fennoscandia. Arct. Alp. Res. 23, 279-286

DELUCIA, E. H. & SMITH, W. K. (1987): Air and soil temperature limitations on photosynthesis in Engelmann spruce during summer. Can. J. For. Res. 17, 527-533

EHRHARDT, F. (1961): Untersuchungen über den Einfluß des Klimas auf die Stickstoffnachlieferung von Waldhumus in verschiedenen Höhenlagen der Tiroler Alpen. Forstwiss. Cbl. 80, 193-215

EMANUELSSON, U. (1987): Human Influences on the vegetation in the Torneträsk area during the last three centuries. Ecol. Bull. 38, 95-111

FRITTS, H. C. (1976): Tree-rings and climate. Academic Press, New York

GOLDSTEIN, G. H. (1981): Ecophysiological and demographic studies of white spruce (*Picea glauca* (Moench) Voss) at treeline in the Central Brooks Range of Alaska. Ph. D. Thesis, Univ. Washington, Seattle

GRACE, J.; ALLEN, S. J. & WILSON, C. (1989): Climate and the meristem temperatures of plant communities near the tree-line. Oecologia 79, 198-204

HANSEN-BRISTOW, K. (1986): Influence of increasing elevation on growth characteristics at timberline. Can. J. Bot. 64, 2517-2523

JACOBY, G. C. (1983): A dendroclimatic study in the forest-tundra ecotone on the shore of Hudson Bay. Nordicana 47, 95-99

KARLSSON, P. S. & NORDELL, K. O. (1988a): Growth of *Betula pubescens* and *Pinus sylvestris* seedlings in a subarctic environment. Funct. Ecol. 1, 37-44

KARLSSON, P. S. & NORDELL, K. O. (1988b): Intraspecific variation in nitrogen status and photosynthetic capacity within mountain birch populations. Holarct. Ecol. 11, 293-297

KÖPPEN, W. (1936): Das geografische System der Klimate. Bornträger, Berlin

KÖRNER, Ch. & RENHARDT, U. (1987): Dry matter partitioning and root length/leaf area ratios in herbaceous perennial plants with diverse altitudinal distribution. Oecologia 74, 411-418

LARCHER, W. (1973): Temperature dependence of other metabolic processes. In: Prech, H.; Christophersen, J.; Hensel, H. & Larcher, W. (eds.): Temperature and life. Springer-Verlag, New York, 137-144

MASUZAWA, T. (1985): Ecological studies on the timberline of Mt. Fuji. I. Structure of plant community and soil development of the timberline. Bot. Mag. Tokyo 98, 15-28

ROHMEDER, E. (1967): Beziehungen zwischen Frucht- bzw. Samenerzeugung und Holzerzeugung der Waldbäume. Allg. Forstzeitschr. 22, 33-39

SCOTT, P. A.; BENTLEY, C. V.; FAYLE, D. C. F. & HANSELL, R. I. (1987): Crown forms and shoot elongation of white spruce at the treeline, Churchill, Manitoba, Canada. Arct. Alp. Res. 19, 175-186

SKRE, O. (1972): High temperature demands for growth and development in Norway spruce (*Picea abies* (L.) Karst.) in Scandinavia. Meddr. Nor. Landbrukshøgskole 51, 1-29

SVEINBJÖRNSSON, B. (1983): Bioclimate and its effect on the carbon dioxide flux of mountain birch (*Betula pubescens* Ehrh.) at its elevational tree-line in the Torneträsk area, Northern Sweden. Nordicana 47, 111-122

SVEINBJÖRNSSON, B. (1992): The arctic treeline in a changing climate. In: Chapin, F. S. III.; Reynolds, J. F.; Jeffries, R. L.; Shaver, G. R.; Svoboda, J. & Chu, E. (eds.): Arctic ecosystems in a changing climate - an ecophysiological perspective: Arctic vegetation in a changing climate. Academic Press San Diego, 239-256

SVEINBJÖRNSSON, B.; NORDELL, O. & KAUHANEN, H. (1992a): Nutrient relations of mountain birch growth at and below the altitudinal tree-line in Swedish Lapland. Funct. Ecol. 6, 213-220

SVEINBJÖRNSSON, B.; SONESSON, M.; NORDELL, O. & KARLSSON, S. (1992b, in press): Mountain birch tree performance in different environments in Sweden and Iceland. In: Alden, J. N. & Mastrantonio, L. (eds.): Forest development in cold climates. Plenum Publishing, New York

TENOW, O. (1975): Topographical dependence of an outbreak of *Oporinia autumnata* Bkh. (Lep., Geometridae) in a mountain birch forest in Northern Sweden. Zoon 3, 85-110

THORSTEINSSON, I. (1986): The effect of grazing on the stability and development of northern rangelands: a case study of Iceland. In: Gudmundsson, Ó. (ed.): Grazing research at northern latitudes. Plenum Publishing, 37-43

TRANQUILLINI, W. (1959): Die Stoffproduktion der Zirbe (*Pinus cembra* L.) an der Waldgrenze während eines Jahres. Planta 54, 107-151

TRANQUILLINI, W. (1979): Physiological ecology of the alpine timberline: tree existence at high elevations with special references to the European Alps. Springer-Verlag, Berlin

TRYON, P. R. & CHAPIN, F. S. III. (1983): Temperature control over root growth and biomass in taiga forest trees. Can. J. For. Res. 13, 827-833

VOWINKEL, T. (1975): The effect of climate on the photosynthesis of *Picea mariana* at the subarctic tree line. Ph. D. Thesis, McGill Univ., Montréal, 147 p.

VOWINCKEL, T.; OECHEL, W. C. & BOLL, W. G. (1975): The effect of climate on the photosynthesis of *Picea mariana* at the subarctic tree-line. Can. J. Bot. 53, 604-620

Address of the author:

Prof. Dr. B. Sveinbjörnsson, University of Alaska, Anchorage, 3211 Providence Drive, Anchorage AK 99508, U.S.A.

Glaciological, sedimentological and palaeobotanical data indicating Holocene climatic change in Northern Fennoscandia

Wibjörn Karlén

Summary

Holocene climatic changes in Fennoscandia are known from changes in vegetation distribution and glacier size fluctuations. In general, the climate became warmer shortly after deglaciation and remained relatively warm up to about 5000 yr B.P. A distinctly cold period lasting from the seventeenth century to the late nineteenth century is well documented. Evidence of short term climatic changes superimposed on this general trend are pointed out and discussed. It is likely that particularly warm periods occurred around 8000, 6800, 6000, 5000, 4500, 3700, 2800, 1500 and 900 yr B.P. Particularly cold and/or wet periods seem to have occurred around 7300, 4700, 3000, 2000, 1100 yr B.P. and during the last 600 years. Less distinct cold fluctuations occurred around 6700, 6300, 5700, 4200 ^{14}C yr B.P. The magnitude of the fluctuations are likely to have been around 2°C.

Zusammenfassung

Die holozänen Klimaschwankungen Fennoskandiens konnten anhand der Veränderungen der Vegetationsausbreitung und der unterschiedlichen Gletscherausdehnungen rekonstruiert werden. Generall läßt sich sagen, daß das Klima kurz nach dem Ende der Vergletscherung wärmer wurde und bis etwa 5000 J.v.h. relativ warm blieb. Eine deutlich kalte Periode vom siebzehnten bis zum neunzehnten Jahrhundert ist gut dokumentiert. Hinweise auf kurzzeitige Klimaschwankungen, die diesen allgemeinen Trend überlagern, werden aufgezeigt und erörtert. Deutlich wärmere Perioden traten höchstwahrscheinlich um 8000, 6800, 6000, 5000, 4500, 3700, 2800, 1500 und 900 J.v.h. auf. Entsprechend kältere und / oder feuchtere Perioden herrschten um 7300, 4700, 3000, 2000, 1100 J.v.h. und in den letzten 600 Jahren. Weniger stark ausgeprägte kalte Schwankungen traten um 6700, 6300, 5700, 4200 ^{14}C-J.v.h. auf. Das Ausmaß der Schwankungen beträgt wahrscheinlich ca. 2°C.

1. Introduction

The existence of Holocene climatic changes is well known in Fennoscandia from studies of fossil pines above the present pine tree limit (LUNDQVIST, 1969; KARLÉN, 1976), recurrence surfaces in raised peat bogs (LUNDQVIST, 1957; BARBER, 1982), glacier moraines located far outside the present-day glacier fronts (e.g. ALEXANDER & WORSLEY, 1973; BERGSTRÖM, 1954; ELVEN, 1978; FAEGRI, 1934, 1950; GRIFFEY & MATTHEWS, 1978; GRIFFEY & WORSLEY, 1978; HOLE & SOLLID, 1979; KARLÉN, 1976, 1982, 1988; MATTHEWS, 1975, 1980; MATTHEWS & SHAKESBY, 1984; WORSLEY, 1974), as well as pollen and macrofossils which show that several plant species have been common much farther north than their present-day northern limit (ANDERSSON, 1902).

In Fennoscandia it is generally accepted that climate became warmer shortly after deglaciation around 10 000 yr B.P. and that a broad "climatic optimum" in mid-Holocene was followed by a declining climate (BERGLUND, 1968). Deviations from a smooth pattern have been suggested, but there is no general agreement about these details. However, the existence of a cold event during the last few hundred years is usually accepted. In this paper evidence indicating short fluctuations superimposed on the general climatic trend will be pointed out.

Table 1 Mean temperature and precipitation for a continental area (Karesuando) and a maritime area (Tromsø)

	Temperature (1931-1960) (°C)		Precipitation (1931-1960) (mm)	
	Karesuando	Tromsø	Karesuando	Tromsø
January	-14,0	-3,5	19	96
February	13,9	4,0	18	79
March	-9,9	-2,7	17	91
April	-3,6	0,3	19	65
May	3,0	4,1	26	61
June	9,8	8,8	46	59
July	13,7	12,4	63	56
August	11,2	11,0	57	80
September	5,7	7,2	41	109
October	-1,6	3,0	26	115
November	-7,3	-0,1	26	88
December	-11,2	-1,9	22	95
Annual	-1,5	2,9	380	994

2. Area

Northern Fennoscandia includes contrasting areas such as the maritime west coast of Norway and the continental areas of Northern Sweden and Finland. In spite of being located north of the Arctic circle, the Atlantic Ocean is open all year. The mean temperature for Tromsø is -4.0°C (1931-1960) in February and 12.4°C in July. In contrast, Karesuando located at the border between Finland and Sweden, has a continental climate with a mean temperature of -13.9°C in February and +13.7°C in July. Temperature observations at stations in Northern Fennoscandia are well correlated. The west coast of Norway receives large amounts of precipitation, but in the area east of the mountains the precipitation is fairly small (Table 1).

3. Glacier evidence of short-term climatic changes

Glacier response to climate is well known (PATERSON, 1981). In the investigated region the mass balance depends on summer temperature and winter precipitation. Frequently, glaciers in continental areas are sensitive to changes in summer temperature whereas glaciers in maritime areas are sensitive to changes in winter precipitation. If a change in temperature or precipitation is substantial, glaciers in a broad zone between the extreme continental and maritime areas will respond to it (NESJE, 1989).

Glacier fronts will respond to changes in mass balance only after some years. It is frequently believed that the time lag is in the order of 50-100 years. However, observations of Norwegian glaciers show that the delay may be as short as a few years for small, steep glaciers (LIESTÖL, 1967). Also, observation of glacier front advances and retreats in the Alps during the twentieth century show that glacier fronts react within a few years (Fig. 1). However, all glaciers do not respond at the same time.

Historical documents facilitate a detailed history of glacier fluctuations in Southwestern Norway (references in GROVE, 1979, 1988; KARLÉN, 1982, 1988). However, information is available for only a few glaciers. The detailed descriptions of the 1748-advance of the glacier Nigaard is well known. It is frequently assumed that most glaciers in Norway reached their Holocene maximum extent at the same time. Historical documents, paintings, photographs, and measurement show that the general retreat in the eighteenth and nineteenth centuries was interrupted by small advances around 1810, 1825, 1835, 1850, 1870, 1890, 1910, and 1925 (KARLÉN, 1988).

A large number of glacier moraines in Norway and a small number in Sweden are ^{14}C dated. The samples are mostly obtained on buried soils but a few dates are obtained on wood from trees overrun during glacier advances (WORSLEY & ALEXANDER, 1975). The dating of soils buried beneath moraines is problematic (MATTHEWS, 1980; MATTHEWS &

DRESSER, 1983) and errors may be large. KARLÉN & DENTON (1975) have tested the technique of dating buried soils by sampling a soil underneath a moraine formed during the early twentieth century and have concluded that the error may be as limited as a few hundred years. Until the technique is tested further, ^{14}C dates on soils buried beneath moraines must be considered uncertain, but the results should not be disregarded.

Results from radiocarbon dating of moraines are summarized by KARLÉN (1988). The dating of buried soils indicates that a few glaciers advanced around or shortly after 6300 yr B.P. and 5000 yr B.P. Glaciers may have advanced to positions as extended as those during the Little Ice Age around 3000, 2400, 2000, and 1300-1100 yr B.P. Some dates from the latter period are obtained on sheared-off tree stumps and are therefore likely to be quite precise. ^{14}C dates on buried soils may indicate that a few glaciers advanced 600-400 yr B.P.

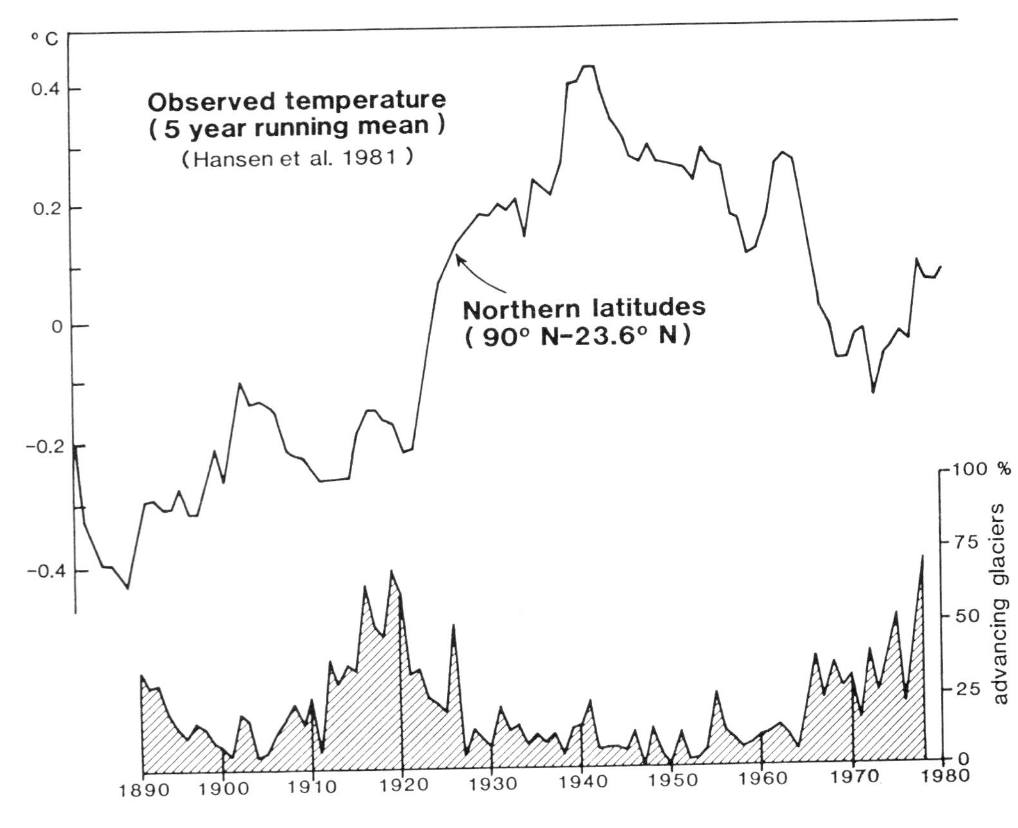

Fig. 1 A glacier advance is a good indication of a decrease in summer temperature or an increase in winter precipitation. Glacier advances in the Alps lag only slightly behind changes in observed temperature of the northern hemisphere

Organic debris suitable for ^{14}C dating is only rarely found in association with moraines and because of large variations in the atmospheric content of ^{14}C (STUIVER, 1982) the technique cannot be used for dating Little Ice Age moraines. A technique, lichenometry, developed by BESCHEL (1961, 1973) can frequently be applied for dating Little Ice Age moraines and occasionally older moraines can also be dated by this technique. The technique is simple but the precision is limited, in particular for old moraines. For calibration of the growth rate the technique depends on independent dating of reference surfaces such as moraines or large rock surfaces. Mining operations during the seventeenth, eighteenth, and nineteenth centuries have left waste heaps suitable for calibration in Sweden (KARLÉN 1973, 1975, 1976). Unfortunately, few calibration surfaces of this type seem to be known in Northern Norway. Attempts to use the lichenometric technique without proper calibration surfaces have led to wide differences in opinions about the time of glacier expansions in Norway (INNES, 1984; BALLANTYNE, 1990).

Lichenometric studies in Southern Norway (MATTHEWS, 1974, 1975, 1976, 1977; HOLE & SOLLID, 1979; ERIKSSTAD & SOLLID, 1986) confirm the historical record and add information about glacier expansions. Evidence of an advance around 1780 has been obtained at a few glaciers (HOLE & SOLLID, 1979).

Beyond 300-400 yr B.P., ^{14}C dated moraines have been used for the calibration of lichenometric dates. Since the quality of these dates has been disputed, the lichenometric dates must be regarded as unprecise. However, from the results obtained it seems safe to conclude that glaciers advanced during several periods in the Holocene. During at least the last of these periods, the Little Ice Age, glaciers reached extended positions on several occasions. The small number of lichenometric dates on moraines in Norway older than the Little Ice Age (MATTHEWS & SHAKESBY, 1984) may indicate that the maritime glaciers did not advance extensively during cold periods when the Swedish, continental type glaciers advanced. This is an indication that major changes occurred in temperature, while the precipitation remained relatively low during the Early and mid-Holocene periods.

Streams coming from active glaciers carry much more silt than other streams in the Scandinavian mountains. This silt transport increases the minerogenic content of proglacial lacustrine sediments during periods when there are active glaciers in a drainage basin. The silt release from a glacier is not understood in detail and there is some debate about the timing of the peak transport in relation to maximum glacier extent. In spite of this, the technique can yield continuous data about glacier advances if a lake is located in an area where slope processes and erosion of the stream bed do not disturb the glacier signal. Major fluctuations in silt content, and hence periods of active glaciers in an area, can be dated reasonably well with the ^{14}C technique.

A sediment core from Vuolep Allakasjaure, a lake below a small, fast responding glacier in Northern Sweden, is believed to depict fluctuations in climate (Fig. 2). Several cores have been retrieved. There is some discrepancy between these cores but the general pattern is similar. The sediment sequence is well dated and the dates are consistent with a close to linear rate of sedimentation, except for during the last millennium, when the rate of sedimentation decreased. The latest 1500 years of the lacustrine sediment record indicates climatic changes similar to the ones shown by a dendrochronological record based on pine tree-ring width, which was obtained in the same area (BARTHOLIN & KARLÉN, 1983; KARLÉN, 1984) (Fig. 3).

According to these sediment cores, advances have occurred around 7300, 4700, 3000, 2000, 1100 ^{14}C yr B.P. and during the last 600 years (KARLÉN, 1981). In addition, small advances seem to have occurred around 6700, 6300, 5700, and 4200 ^{14}C yr B.P.

Sediment cores from other proglacial lakes in Northern Sweden show that several glaciers melted away completely after deglaciation and reformed in 3000-2000 yr B.P. (KARLÉN, 1981) (Fig. 4).

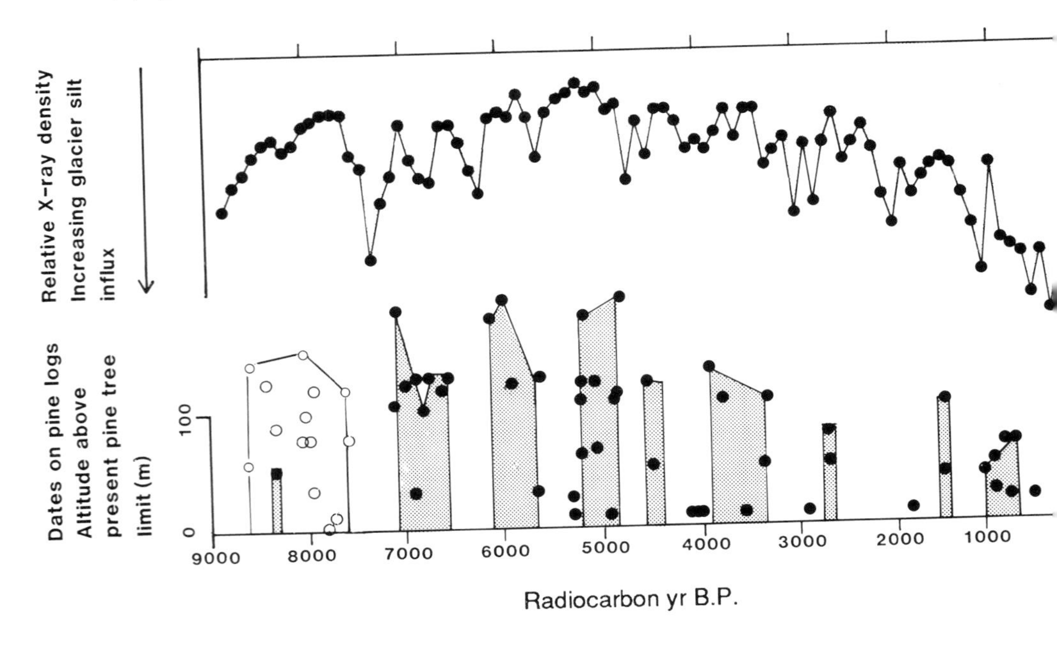

Fig. 2 The upper diagram shows relative variations in the silt content of a proglacial lake called Vuolep Allakasjaure. A small glacier in the drainage basin is believed to respond to changes in summer temperature and possibly also winter precipitation. The lower diagram shows ^{14}C dates obtained from pine logs found above the present pine tree limit in Northern Sweden (black dots). Only a few ^{14}C dates from fossil pine from Northern Sweden have been obtained so far from the period around 8000 ^{14}C yr B.P. The diagram is therefore complemented by dates from the Early Holocene obtained from South Central Sweden (open circles)

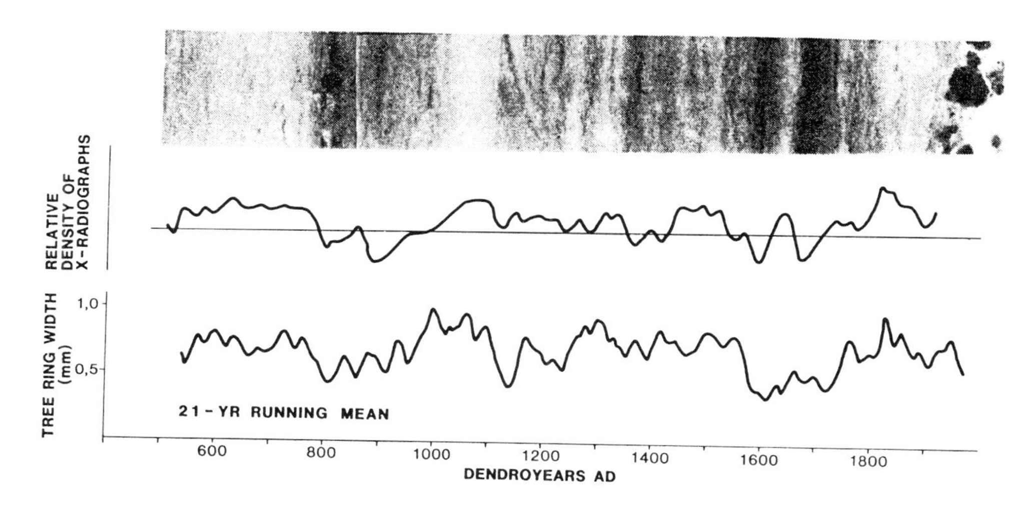

Fig. 3 Relative X-ray density of the upper 10 cm of a sediment core from a pro-glacier lake and variations in mean tree-ring width of pines from Northern Sweden show the same fluctuations in climate (BARTHOLIN & KARLÉN, 1983; KARLÉN, 1984)

4. Fossil pine above the present pine tree limit

Dates on fossil pine reveal information about warm periods. In contrast to the methods to date glacier advances, samples of fossil pine yield good ^{14}C dates. The existence of pine trees that have germinated above the present climatic boundary proves that the climate was favourable during at least some parts of the Holocene. A lack of evidence cannot in itself prove the existence of cold events between dated periods of high pine tree limits.

Fossil pines found above the present pine tree limit are dated by the ^{14}C technique (references in KARLÉN, 1976). A large number of dates have also been obtained in the mountains in South Central Sweden (LUNDQVIST, 1969) and recently more dates have been obtained from this area by KULLMAN (1980, 1987, 1988). The occurrence of pine above the present pine tree limit in Norway has recently been discussed by MOE (1979), BARTH et al. (1980), HAFSTEN (1981), and AAS & FAARLUND (1988). In Finland, evidence of a pine forest covering a larger area than the present one has been discussed by ERONEN (1979). The altitude of the local pine tree limit varies because of local climatic condition. In diagrams in which the dates are shown in relation to altitude above sea level there is no distinct pattern of periods with a concentration of dates on pine. A pattern only becomes obvious when the altitude of the dated samples is shown in relation to the present pine tree limit (KARLÉN, 1976; KULLMAN, 1987). To some extent the dates are clustered. A limited dispersion of the dates may be a result of problems in determining present pine tree line altitude, errors in the

^{14}C technique (often 100-200 yr), and the long life span of pine (BRIFFA et al., 1990). In addition, a few pines may well have managed to become established above the present pine tree limit during short warm intervals, which most likely occurred even during generally cold centuries. The dates of fossil pine may lag somewhat behind the periods of warm summer climate that are favourable for germination at high altitude.

The ^{14}C dates of pine from above the present pine tree limit in Northern Sweden indicate that pine reached its maximum altitude around 5000 ^{14}C yr B.P. and then decreased. In the mountains of South Central Sweden the maximum extent of the pine tree limit occurred around 8000 yr B.P. (Fig. 2). The few dates on pine in Northern Sweden from this time may indicate that pine migrated into this area later than it did into Northern Finland, South Central Sweden, and Norway (LUNDQVIST, 1969; HYVÄRINEN, 1975).

Clusters of dates from pine logs found above the present pine tree limit around 6800, 6000, 5000, 4500, 3700, 2800, 1500, and 900 yr B.P. indicate warm summer climate at approximately these times in Northern Sweden. As mentioned above, in the South Central Swedish mountains the maximum pine expansion occurred around 8000 yr B.P. To some extent the high altitude to which pine spread at this period depended on the land rebound.

KULLMAN (1987) has reported dates around 8000, 7000, 6000, 5000, and 1000 ^{14}C yr B.P. from pine logs found above the present pine tree limit in the South Central Swedish mountains. In addition to these groups of dates several other dates have been reported. In general the ages of these dated logs coincide with dates from Northern Sweden.

MOE (1979) found a maximum in pine distribution in Southern Norway around 8000 yr B.P. and a temporary maximum around 5000 yr B.P. From a diagram in AAS & FAARLUND (1988) it seems likely that pine germinated at relatively high altitude in Southern Norway around 8000, 7000, and 6000 yr B.P. These dates also coincide with periods of a high pine tree limit in Northern Sweden.

The possibility of the occurrence of tree limit variations with a duration of only a few hundred to a thousand years is not generally accepted. However, evidence concerning expansion and retreat of the pine tree limit within a short period of time is available.

A small expansion was observed during the 1930s (HUSTICH, 1948, 1958). Seedlings were established above the pine tree limit and these plants were still growing in the 1950s. The cold periods discussed here are generally over 500 years long and therefore longer than the typical life-span of 300 years observed in dendrochronological studies (BRIFFA et al., 1990). Also, a process which can remove stands within only a few years was observed recently when large stands of spruce at the present tree limit were damaged after one cold summer (KULLMAN, 1989). KULLMAN (pers. comm. 1990) has pointed out that pine respond similarly to cold climate.

5. Palynological data

Palynology has a long tradition in Fennoscandia. The results can only be discussed very briefly here. The relative frequency of pollen taxa in lacustrine sediments and peat bogs have been used primarily for determining vegetation history. Frequently, the conclusions are limited to a determination of pollen zones; to some extent, the boundaries depend on changes in climate although migration also affects these zone boundaries.

A widely used palynological study depicting the climate has been published by BERGLUND (1968). This mainly describes conditions in Southern Sweden and Denmark but it does illustrate the accepted view of the Holocene climate in Scandinavia. Around 8000 B.C. a temperature increase began. During the Early Holocene the increase in temperature was rapid. The rise continued to around 4000 B.C. A smooth decrease in temperature of 2°C followed. During the last 3000 years the temperature has oscillated slightly.

A view of the climate covering the last 6000 years, which is similar to BERGLUND's, has been given by RENBERG & SEGERSTRÖM (1981). The information is based on pollen and sedimentological studies of varved sediments from a lake in Northern Sweden. A temperature maximum was reached close to 3000 B.C, and two cold periods are ^{14}C dated to 500 B.C. and 1600 A.D. A period of increased spring runoff is believed to have occurred around 0 A.D.

HYVÄRINEN (1975) has constructed absolute and relative pollen diagrams for three lakes in Northern Finland. The results show that pine had immigrated into Northern Finland as early as 8500 yr B.P. and at that time extended farther north than it does at present. Pine remained at its maximum extent from 7500 yr B.P. to 5000 yr B.P., when the pine recession started. The pine tree limit retreat continued to 3000-2500 yr B.P. During the last several thousand years little change in the pine tree limit has been noticed in the pollen record.

The annual mean influx of pollen in a small lake near the coast of Northern Sweden has been studied (SEGERSTRÖM, 1990). The study was facilitated by the occurrence of distinct annual varves in the sediments. Particularly the influx of *Betula* and *Alnus* has varied considerably during the slightly more than 6000 years that have passed since the lake was formed. It is believed that the variations in pollen influx in the sediments reflect climatic change. Several periods of relatively large influx of pollen are dated by the counting of varves to around 3500-3200, 2600-2100, 1700, 1000, 400 B.C. and 600, 1050, and 1900 A.D. (4700-4400, 4000-3600, 3300, 2800, 2300, 1400, 1000 ^{14}C yr B.P. and the recent century; cp. Fig. 5).

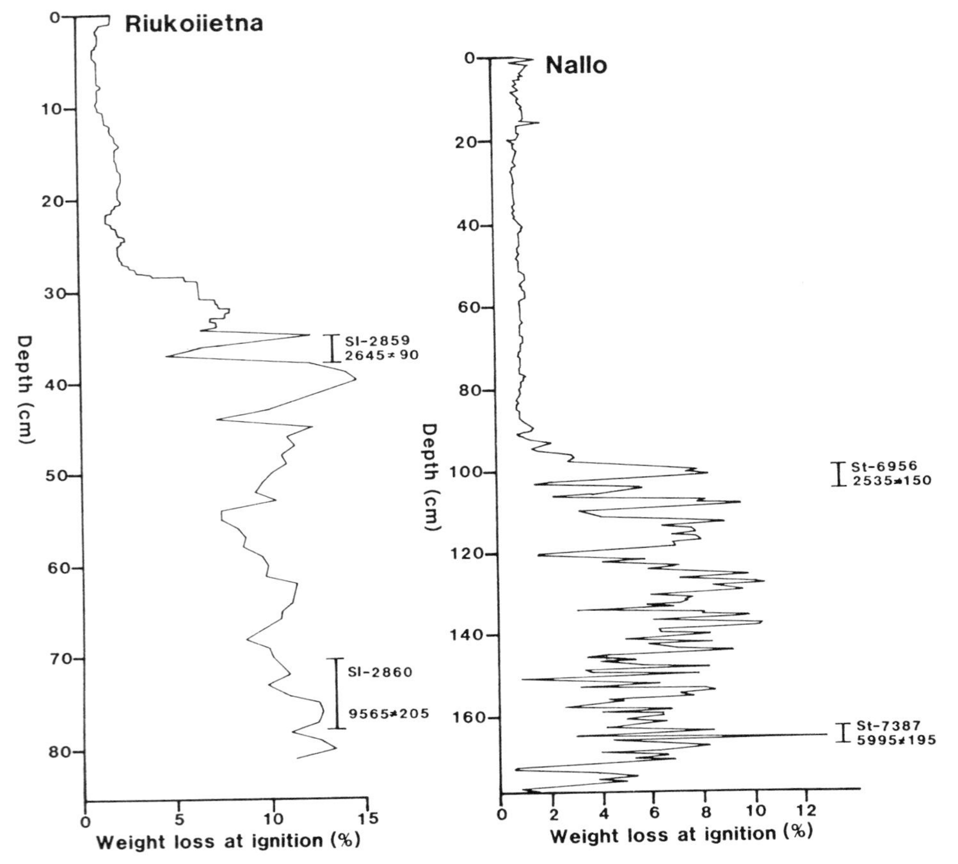

Fig. 4 Weight loss at ignition obtained on sediments from proglacial lakes located in the Kebnekaise massif, Northern Sweden. A) Before approx. 2500 yr B.P. a lake near Nallo received glacier melt water intermittently. After this date, glacier melt water reached the lake continuously. Apparently, the small glacier in the drainage basin increased distinctly in size around 2500 yr B.P. The outermost moraines bordering the unnamed glacier have been dated by lichenometry to about 2500 yr B.P. (KARLÉN, 1975, 1976). B) Apparently the glacier Riukojietna melted away during deglaciation and did not reform until about 2600 ^{14}C yr B.P (KARLÉN, 1981)

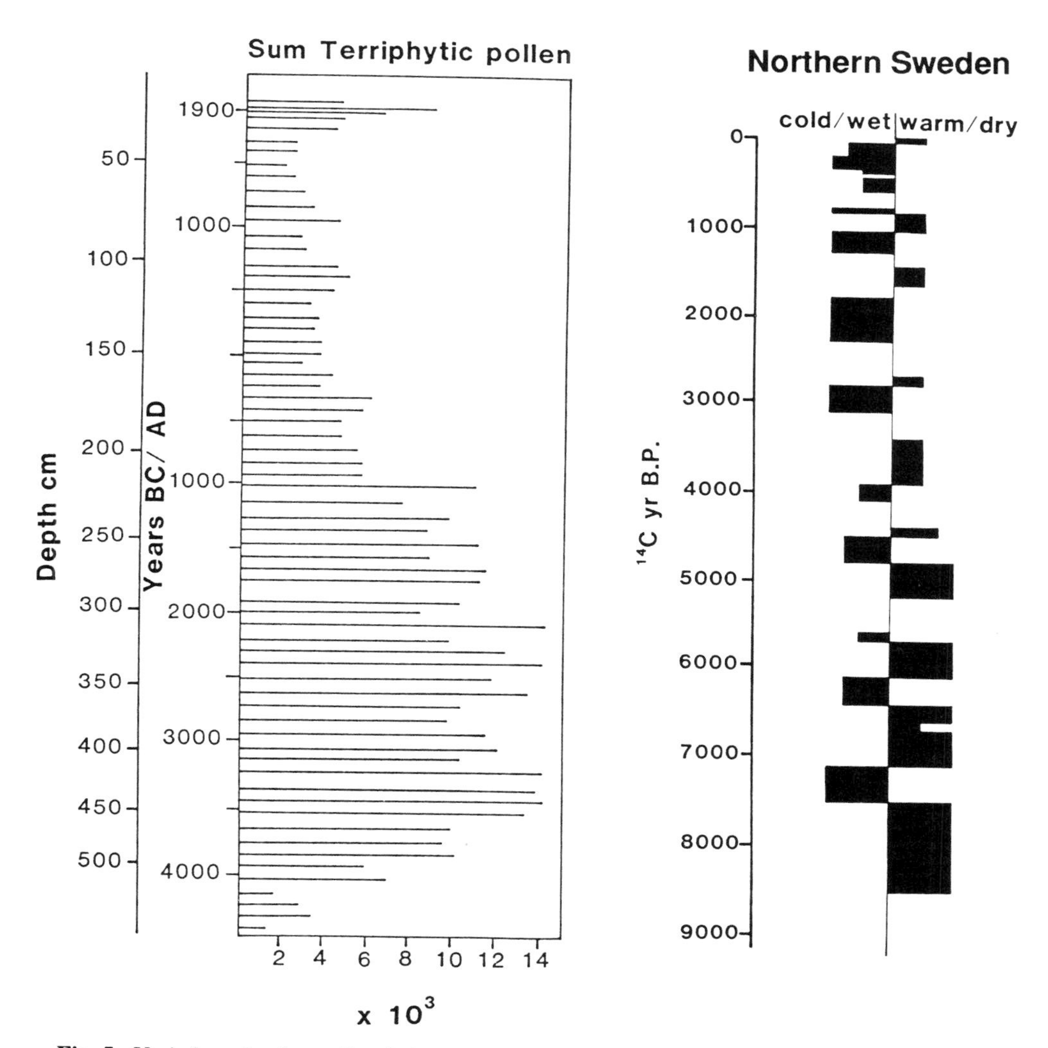

Fig. 5 Variations in the pollen influx in a small lake with varved sediments at the coast of Northern Sweden. Note that the age is calculated on the basis of varve counts and therefore the time scale is in calendar B.C/A.D. Greatly increased pollen influx coincides with periods of high tree limit (from SEGERSTRÖM, 1990)

Fig. 6 Generalized stratigraphy for Northern Scandinavia. Information from proglacial lacustrine sediments, pine tree limit variations, and ^{14}C dated moraines from Scandinavia have been used for the construction of the diagram

6. Discussion and conclusions

It is generally believed that the temperature rose after deglaciation and that a maximum was reached at approximately 5000-6000 yr B.P. Since this "Hypsithermal" period the temperature has decreased. This pattern is known from studies of pollen in peat and lacustrine sediments, the altitude of the pine tree limit, and glacier fluctuations.

It is accepted by a broad group of scientists that climate deviated from the smooth pattern described above during the Little Ice Age and also that it may have oscillated during the last 3000 years. The evidence comes basically from the same sources as mentioned above.

A few scientists believe that the climate has fluctuated distinctly during the entire Holocene. Evidence for this comes from well dated proglacial lacustrine sediments, which yield a continuous record of how a small glacier responds to climate.

The general pattern is largely in phase with pollen influx data from the last 6000 years dated by varve counting as well as with ^{14}C dates on variations in the pine tree limit.

The amplitude of the Holocene climatic changes permitted small, fast-responding glaciers to reach extended positions even during the Hypsithermal period. Some glaciers are known to have formed first after 3000 yr B.P. when the climate deteriorated markedly. Large glaciers may have responded with only limited advances and moraines were possibly formed inside moraines of the Little Ice Age. These moraines were mostly overrun and destroyed by extensive Late Holocene glacier advances. However, a few moraines from Early and mid-Holocene glacier advances, protected from later overrunning, have survived. A generalized view of Holocene climatic changes in northern Scandinavia is given in Fig. 6.

Acknowledgements

The fieldwork on which this study is based has been supported by the Swedish Natural Science Research Council.

References

AAS, B. & FAARLUND, T. (1988): Postglaciale skoggrenser i sentrale sörnorske fjelltrakter. ^{14}C-datering av subfossile furu- og björkerester. Norsk Geogr. Tidsskr. 42, 25-61

ALEXANDER, M. J. & WORSLEY, P. (1973): Stratigraphy of a Neoglacial end moraine in Norway. Boreas 2,117-142. Oslo

ANDERSSON, G. (1902): Hasseln i Sverige fordom och nu. Sver. Geol. Unders., Serie Ca. 3, 162 p.

BALLANTYNE, C. K. (1990): The Holocene glacial history of Lyngshalvöya, Northern Norway: chronology and climatic implications. Boreas 19, 93-117. Oslo

BARBER, K. E. (1982): Peat-bog stratigraphy as a proxy climate record. In: Harding, A. (ed.): Climatic change in later pre-history. Edinburgh Univ. Press, 103-113

BARTH, E. K.; LIMA-DE-FARIA, A. & BERGLUND, B. E. (1980): Two ^{14}C dates of wood samples from Rondane, Norway. Bot. Notiser 133, 643-644

BARTHOLIN, T. S. & KARLÉN, W. (1983): Dendrokronologi i Lappland A.D. 436-1981. Dendrokronologiska Sällskapet, Meddelande 5, 3-16

BERGLUND, B. (1968): Vegetationsutvecklingen i Norden efter istiden. Sveriges Natur 1968, 31-52

BERGSTRÖM, E. (1954): Studies of the variations in size of Swedish glaciers in recent centuries. AIHS, Assemblée Générale de Rome, Publ. No. 39, 356-366

BESCHEL, R. E. (1961): Dating rock surfaces by lichen growth and its application in glaciology and physiography (lichenometry). In: Raasch, G. O. (ed.): Geology of the Arctic. Univ. of Toronto Press, Toronto, 1044-1062

BESCHEL, R. E. (1973): Lichens as a measure of the age of recent moraines. Arct. Alp. Res. 5/4, 303-309

BRIFFA, K. R.; BARTHOLIN, T. S.; ECKSTEIN, D.; JONES, P. D.; KARLÉN, W.; SCHWEINGRUBER, F. H. & ZETTERBERG, P. (1990): A 1,400 year tree-ring record of summer temperatures in Fennoscandia. Nature 346 (2 Aug), 434-439

ELVEN, R. (1978): Subglacial plant remains from Omnsbreen glacier area, South Norway. Boreas 1, 83-89

ERIKSTAD, L. & SOLLID, J. L. (1986): Neoglaciation in south Norway using lichenometric methods. Norsk Geogr. Tidsskr. 40, 85-105

ERONEN, M. (1979): The retreat of pine forest in Finish Lappland since the Holocene climatic optimum: a general discussion with radiocarbon evidence from subfossil pines. Fennia 157/2, 93-114

FAEGRI, K. (1934): Uber die Längenvariationen einiger Gletscher des Jostedalsbre und die dadurch bedingten Pflanzensukzessionen. Bergens Museums Årbok 7/2 (1933), 1-255

FAEGRI, K. (1950): On the variations of Western Norwegian glaciers during the last 200 years. Assemblée Générale d'Oslo de l'Union Géodésique et Géophysique Internationale, 293-303

GRIFFEY, N. J. & MATTHEWS, J. A. (1978): Major Neoglacial glacier expansion episodes in Southern Norway: Evidences from moraine ridge stratigraphy with ^{14}C dates on buried palaeosols and moss layers. Geogr. Ann. 60A(1-2),73-90

GRIFFEY, N. J. & WORSLEY, P. (1978): The pattern of Neoglacial glacier variations in the Okstindan region of Northern Norway during the last three millenia. Boreas 7, 1-17

GROVE, J. M. (1979): The glacial history of the Holocene. Progr. Phys. Geogr. 3/1, 1-50

GROVE, J. M. (1988): The Little Ice Age. Methuen, 498 p.

HAFSTEN, U. (1981): An 8000 years old pine trunk from Dovre, South Norway. Norsk Geogr. Tidsskr. 35, 161-165

HANSEN, J.; JOHNSON, D.; LACIS, A.; LEBEDEFF, S; LEE, P.; RIND, D. & RUSSELL, G. (1981): Climatic impact of increasing atmospheric carbon dioxide. Science 213, 957-965

HOLE, N. & SOLLID, J. L. (1979): Neoglaciation in Western Norway - preliminary results. Norsk Geogr. Tidsskr. 33, 213-215

HUSTICH, I. (1948): The Scotch pine in northernnmost Finland and its dependence on the climate in the last decades. Acta Bot. Fenn. 42, 3-75

HUSTICH, I. (1958): On the recent expansion of the Scotch pine in Northern Europe. Fennia 82/3, 1-25

HYVÄRINEN, H. (1975): Absolute and relative pollen diagrams from northernmost Fennoscandia. Fennia 142, 1-23

INNES, J. L. (1984): Lichenometric dating of moraine ridges in Northern Norway: some problems of application. Geogr. Ann. 66A/4,341-352

KARLÉN, W. (1973): Holocene glacier and climatic variations, Kebnekaise mountains, Swedish Lappland. Geogr. Ann. 55A/1, 29-63

KARLÉN, W. (1975): Lichenometrisk datering i norra Skandinavien - metodens tillförlitlighet och regionala tillämpning. Department of Physical Geography, Univ. of Stockholm, Report 22, 67 p.

KARLÉN, W.(1976): Lacustrine sediments and tree-limit variations as indicators of Holocene climatic fluctuations in Lappland: Northern Sweden. Geogr. Ann 58A/12, 1-34

KARLÉN, W. (1981): Lacustrine sediment studies. Geogr. Ann. 63A/3-4, 273-281

KARLÉN, W. (1982): Holocene glacier fluctuations in Scandinavia. Striae 18, 26-34, Uppsala

KARLÉN, W. (1984): Dendrochronology, mass balance and glacier front fluctuations in Northern Sweden. In: Mörner, N.-A. and Karlén, W. (eds.): Climatic change on a yearly to millennial basis. D. Reidel Publishing Company, 263-271

KARLÉN, W. (1988): Scandinavian glacier and climatic fluctuations during the Holocene. Quaternary Science Reviews 7,199-209

KARLÉN, W. & DENTON, G.H. (1975): Holocene glacier variations in Sarek National Park, Northern Sweden. Boreas 5, 25-56

KULLMAN, L. (1980): Radiocarbon dating of subfossil Scots pine (*Pinus sylvestris* L.) in the Southern Swedish Scandes. Boreas 9, 101-106

KULLMAN, L. (1987): Sequences of Holocene forest history in the Scandes, inferred from megafossil *Pinus sylvestris*. Boreas 16, 21-26

KULLMAN, L. (1988): Holocene history of the forest-alpine tundra ecotone in the Scandes Mountains (Central Sweden). New Phytol. 108, 101-110

KULLMAN, L. (1989): Cold-induced dieback of montane spruce forests in the Swedish Scandes - a modern analogue of paleoenvironmental processes. New Phytol. 113, 377-389

LIESTØL, O. (1967): Storbreen glacier in Jotunheimen, Norway. Norsk Polarinstitutt Skrifter 141, 1-63

LUNDQVIST, J. (1957): C^{14}-dateringar av rekurensytor i Värmland. Sver. Geol. Unders. Ser. C, 554, 22 p.

LUNDQVIST, J. (1969): Beskrivning till jordartskarta över Jämtlands Län. Sver. Geol. Unders. Ser. Ca. 45, 418 p.

MATTHEWS, J. A. (1974): Families of lichenometric dating curves from Storbreen gletschervorfeld, Jotunheimen, Norway. Norsk Geogr. Tidsskr. 28, 215-235

MATTHEWS, J. A. (1975): Experiments on the reproducibility and reliability of lichenometric dates, Storbreen gletschervorfeld, Jotunheimen, Norway. Norsk Geogr. Tidsskr. 29, 97-109

MATTHEWS, J. A. (1976): "Little Ice Age" palaeotemperatures from high altitude tree growth in S. Norway. Nature 264, 243-245

MATTHEWS, J. A. (1977): A lichenometric test of the 1750 endmoraine hypothesis: Storbreen gletschervorfeld, Southern Norway. Norsk Geogr. Tidsskr. 31, 129-136

MATTHEWS, J. A. (1980): Some problems and implications of ^{14}C dates from a podzol buried beneath an end moraine at Haugabreen, Southern Norway. Geogr. Ann. 62A/3-4, 185-208

MATTHEWS, J. A. & DRESSER, P. Q. (1983): Intensive ^{14}C dating of a buried palaeosol horizon. Geol. Fören Stockh. Förh. 105/1, 59-63

MATTHEWS, J. A. & SHAKESBY, R. A. (1984): The status of the "Little Ice Age" in Southern Norway: relative-age dating of Neoglacial moraines with Schmidt hammer and lichenometry. Boreas 13, 333-346

MOE, D. (1979): Tredgrense-fluktuasjoner på Hardangervidda etter siste istid. In: Nydal, R.; Westin, S.; Hafsten, U. & Gulliksen, S. (eds.): Fortiden i søkelyset, Datering med ^{14}C metoden gjenom 25 år. Trondheim, 199-208

NESJE, A. (1989): Glacier-front variations of outlet glaciers from Jostedalsbreen and climate in the Jostedalsbre region of Western Norway in the period 1901-80. Norsk Geogr. Tidsskr. 43, 3-17

PATERSON, W. S. B. (1981): The physics of glaciers. Pergamon Press, Oxford, 380 p.

RENBERG, I. & SEGERSTRÖM, U. (1981): Varved lake sediments in Northern Sweden. Poster paper, 3rd Int. Symp. on Paleolimnology, 1-8 Sept., Finland

SEGERSTRÖM. U. (1990): The vegetation and agricultural history of a Northern Swedish catchment studied by analysis of varved lake sediments. Unpubl. manuscript. Department of Ecological Botany, Univ. of Umeå, 33 p.

STUIVER, M. (1982): A high-precision calibration of A.D. radicarbon time scale. Radiocarbon 24/1, 1-26

WORSLEY, P. (1974): Absolute dating of the Subboreal climatic deterioration - fossil pine evidence from Strimasund, Väster botten Country, Sweden. Geol. Fören. Stockh. Förh. 96, 399-403

WORSLEY, P. & ALEXANDER, M.J. (1975): Neoglacial palaeoenvironmental change at Engabreen, Svartisen Holandsfjorden, North Norway. Norges Geol. Unders. 321, 37-66

Address of the author:

Prof. Dr. W. Karlén, Department of Physical Geography, Stockholm University, S-106 91 Stockholm, Sweden

Deposits indicative of Holocene climatic fluctuations in the timberline areas of Northern Europe: some physical proxy data sources and research approaches

John A. Matthews

Summary

Physical proxy data indicative of Holocene climatic fluctuations near timberline in Northern Europe are reviewed with emphasis on research from Southern Norway. Approaches based on moraine sequences and moraine stratigraphy, including ^{14}C dates on buried palaeosols, have been widely applied and indicate a Late Neoglacial ("Little Ice Age") glacier maximum in Southern Norway and some parts of maritime Northern Scandinavia. Emerging regional differences in the scale of the "Little Ice Age" glacier expansion episode, relative to earlier Holocene glacier expansion episodes, may be resolved with reference to geographical location in relation to migrating oceanic polar fronts, winter precipitation patterns and continentality. Glacial lake sediments offer the possibility of a continuous record of Holocene glacier variations in regions where Late Neoglacial advances were relatively large. Glacio-fluvial deposits, certain types of periglacial deposit and surface soils are also yielding significant palaeoclimatic information.

Zusammenfassung

In diesem Beitrag werden physische Proxydaten besprochen, die über die holozänen Klimaschwankungen nahe der Waldgrenze in Nordeuropa Auskunft geben können, wobei die Ergebnisse aus Südnorwegen besondere Berücksichtigung erfahren. Forschungsansätze, die unter Einbeziehung ^{14}C-datierter Paläoböden auf der Untersuchung von Moränenstaffeln und der Moränenstratigraphie beruhen, implizieren für Südnorwegen und die maritimen Regionen Nordskandinaviens ein glaziales Maximum im späten Neoglazial ("Kleine Eiszeit"). Auftretende regionale Unterschiede in der zeitlichen Abfolge der Gletschervorstöße während der "Kleinen Eiszeit" können vermutlich, analog zu früheren holozänen Vorstoßphasen, auf die Lage der jeweiligen Region in Abhängigkeit von der Polarfront, der winterlichen Niederschlagsverteilung und der Kontinentalität zurückgeführt werden. In Regionen mit einem vergleichsweise gehäuften Vorkommen von spätneoglazialen Gletschervorstößen besteht die Möglichkeit, mit Hilfe glazigener limnischer Sedimente für das Holozän eine kontinuierliche Abfolge glazialer Vorstoß- und Rückzugsphasen zu rekonstruieren. Darüber hinaus liefern glazio-fluviale Sedimente, bestimmte

Typen periglazialer Ablagerungen und Bodenoberflächen wichtige paläoklimatische Informationen.

1. Introduction

Approaches to Holocene tree limits and climatic variations based on biological evidence have a long history in Northern Europe. These include micro- and macro-fossil analysis of peats and lake sediments, investigation of the growth and humification of ombrogenous mires, consideration of the distribution of megafossil finds (particularly *Pinus* trunks), and studies involving dendrochronology and dendroclimatology. Several examples of these approaches have been given elsewhere in these workshop proceedings. It is important to realize, however, that there are a number of other approaches that involve largely physical evidence, such as glacial and periglacial deposits, sediments and soils.

Physical proxy evidence is important for several reasons. First, physical data are particularly valuable where biological data are sparse or absent above and beyond alpine and polar tree limits. Second, where physical rather than biological phenomena are involved, a more direct relationship to climate might be expected. It should therefore be possible to obtain a clearer climatic signal from certain types of physical proxy data. Third, climatic effects on physical phenomena may be less susceptible to anthropogenic modification. Certain types of physical evidence are therefore likely to aid the separation of natural from human induced causes of biological change. Fourth, it may be possible to correlate physical evidence with biological responses directly. If so, physical evidence can be used as a substitute for, or at least as a basis for inference about, biological changes.

Most progress in palaeoclimatic reconstruction from physical proxy data near tree limits in Northern Europe, has involved glacier variations. I therefore give some emphasis to this topic, and to my own research in Southern Norway. However, given that several papers in these proceedings address glacier variations, I use this as a justification for concentrating on particular aspects and for introducing some other types of physical proxy data.

2. Glacial deposits

2.1 Moraine sequences and moraine stratigraphy

Variations in the size of glaciers, as indicated by moraine-ridges or till-sheets, have been widely studied. These studies suggest major regional differences in Holocene glacier and climatic variations.

In Southern Norway, stratigraphic and ^{14}C dating studies conclusively demonstrate a Late Neoglacial ("Little Ice Age") glacier maximum (MATTHEWS, 1989, 1990, 1991). Outermost

Neoglacial moraines in front of several outlets of the Jostedalsbreen ice cap are dated by reliable historical evidence to about 1750 A.D. (GROVE, 1985; BØGEN et al., 1989). Supposed earlier glacier maxima within the "Little Ice Age" interval are not supported by reliable historical evidence. Lichenometric dating and weathering-based relative age assessment using the Schmidt hammer suggest that only a small minority (<10%) of glaciers in Southern Norway experienced major pre-"Little Ice Age" Neoglacial advances that exceeded those of the "Little Ice Age" (MATTHEWS & SHAKESBY, 1984; see also ERIKSTAD & SOLLID, 1986). However, the small number of possible older Neoglacial moraines have not been dated, even approximately (cp. ØSTREM, 1965; MCCARROLL, 1989, 1991).

Although certain ^{14}C dates from soils beneath outermost Neoglacial moraines in Jotunheimen were at first interpreted as indicating pre-"Little Ice Age" glacier maxima (GRIFFEY & MATTHEWS, 1978), they have since been reinterpreted in the light of new evidence as reflecting large apparent mean residence times (AMRT's) for carbon in well-developed soils (MATTHEWS, 1980, 1982). Critical evaluation of moraine stratigraphic evidence, with ^{14}C dates from buried soils and plant remains, has established that four Southern Norwegian glaciers - Haugabreen in the Jostedalsbreen region; Vestre Memurubreen, Storbreen and Böverbreen in Jotunheimen - were larger in the "Little Ice Age" than at any other time during Neoglaciation (i.e. during at least the last ca. 5000 years) (MATTHEWS & DRESSER, 1983; MATTHEWS, 1984, 1989, 1990, 1991; MATTHEWS et al., 1986; MATTHEWS & CASELDINE, 1987).

In view of the importance of this evidence for Scandinavian glacier variations, I cite results from one of these sites. Intensive ^{14}C dating of the surface organic (O or FH) horizon of a well-developed and well-preserved Humo-ferric Podzol from beneath the outermost Neoglacial moraine of Haugabreen yielded a significant, near-linear increase in age with depth. Dates as young as 485±55 ^{14}C yr B.P. (CAR-517C) were obtained near the former surface of the buried soil, increasing to 4020±70 ^{14}C yr B.P. (CAR-523C) at 13.0-14.0 cm deeper in the horizon (MATTHEWS & DRESSER, 1983; MATTHEWS, 1984). A further date of 6470±80 ^{14}C yr B.P. (CAR-752C) has since been obtained from an adjacent peat bog. This last date, which extends the record back into the Early Holocene, confirms the likelihood that the glacier was larger in the "Little Ice Age" than at any time since regional deglaciation in the Preboreal (ca. 9000 ^{14}C yr B.P.). Important work from A. Nesje, M. Kvamme and others at the University of Bergen indicates that most Southern Norwegian glaciers probably disappeared in the Early Holocene and reappeared later at various times to fluctuate within their "Little Ice Age" limits (KVAMME, these proceedings; NESJE et al., 1991).

In Northern Sweden, similar studies have led to a rather different model, with the apparent absence of a dominant mid-eighteenth century "Little Ice Age" maximum, lichenometric evidence for earlier "Little Ice Age" glacier advances (but see INNES, 1984)), and major glacier advances throughout the Holocene (KARLÉN, 1973, 1975, 1982, 1988; KARLÉN & DENTON, 1975). In comparison with Southern Norway, the data from Northern Sweden show that many more glaciers exhibit pre-"Little Ice Age" maxima, some as early as 7500 ^{14}C yr B.P. (KARLÉN, these proceedings).

Despite differences in approach and disagreements on details, there can be no doubt that real differences exist in the glacial history of Southern Norway and Northern Sweden. However, these differences do not necessarily apply in a simple way to adjacent regions. Indeed, research from the Lyngen peninsula in the extreme north of Norway appears to agree quite well with the Southern Norwegian model. Pre-"Little Ice Age" Neoglacial maxima are uncommon, most large glaciers seem to have been characterized by a late Neoglacial maximum culminating in the mid-eighteenth century, and small glaciers reached their maxima in the period 1910-1930 A.D. (BALLANTYNE, 1990).

An important advance in the climatic reconstruction of "Little Ice Age" conditions in Lyngen (BALLANTYNE, 1990), suggests that the mid-eighteenth century maximum of the larger glaciers reflects a relatively long period of low winter temperatures with low winter snowfall. At this time, the North Atlantic oceanic polar front was further south, adjacent to Southern Norway (LAMB, 1979). By 1910-1930, on the other hand, the polar front had retreated northwards together with associated depression tracks, bringing greater snowfall to Lyngen than to Southern Norway. As this climatic deterioration was of relatively short duration, only the smaller glaciers were able to attain their Neoglacial maxima at this time. According to this argument, the absence of such a marked glacier advance in 1910-1930 A.D. in Northern Sweden reflects the relatively continental location of the Swedish glaciers, which are less sensitive to varying amounts of winter precipitation (BALLANTYNE, 1990).

A related explanation seems appropriate to account for the importance of the "Little Ice Age" glacier advance relative to earlier Neoglacial advances in Southern Norway. Relatively large "Little Ice Age" climatic deteriorations, which also appear to characterize several other regions (e.g. Iceland, Southwest Alaska and West Greenland), are probably related to geographical location (continentality as well as latitude) in relation to migrating polar fronts and persistent weather patterns. The "Little Ice Age" glacier expansion was relatively large in Southern Norway, when the oceanic polar front was farthest south. Earlier Neoglacial glacier expansion episodes, on the other hand, tend to have been relatively large in Northern Scandinavia (KARLÉN, 1988), when the polar front migrated less far to the south. This latter tendency would have been most marked in continental Northern Sweden, which is farthest from sources of winter snowfall to the west; in maritime Lyngen, three glaciers only provide evidence of pre-"Little Ice Age" Neoglacial advances (BALLANTYNE, 1990). It would appear to be no coincidence that the few examples of possible pre-"Little Ice Age" Neoglacial moraines that have been identified in Jotunheimen are located in the extreme east of the region, the most continental part. This explanation seems also to be consistent with data on Holocene glacier variations from the Svartisen/Okstindan/Saltfjell region of Northern Norway in the vicinity of the Arctic Circle (ALEXANDER & WORSLEY,

1976; GRIFFEY, 1976; WORSLEY & ALEXANDER, 1976; GRIFFEY & WORSLEY, 1978; KARLÉN, 1979), where there is a tendency for the "Little Ice Age" event to have been relatively more important in the maritime west than in the continental east.

2.2 Other approaches based on glacigenic deposits

Despite the progress made in the reconstruction of glacier variations from moraine stratigraphic evidence, the approach is limited by the tendency of any relatively large-scale glacier advances to remove the evidence of older, less-extensive advances. This has led to the analysis of the sedimentary record in glacial lakes, an approach pioneered in Northern Europe by KARLÉN (1976, 1981). Glacier expansion tends to increase the mineral input into glacial lakes (lakes that receive meltwater from modern glaciers), whereas the organic content increases when glacial activity in the catchment is low or absent. Although there are many complicating factors, the best results have been obtained from small, deep, flat-bottomed lakes, which have other lakes upstream (traps for coarse sediment) and which are unaffected by snow avalanches. Under optimum conditions this approach has the potential to produce a continuous sedimentary sequence and hence a continuous record of Holocene glacier variations, including the smaller fluctuations. The completeness of the record of Holocene glacier variations from Northern Sweden owes much to this approach (see KARLÉN, these proceedings).

Similar studies have been made at Vanndalsvatnet in Southern Norway, where the western part of the Spørteggbreen ice cap (between Jostedalsbreen and Jotunheimen) is inferred to have re-formed since 2000 ^{14}C yr B.P. (NESJE et al., 1991) after being absent throughout the Early and mid-Holocene. Many more glacial lakes have been sampled in Southern Norway and are in earlier stages of investigation (KARLÉN & MATTHEWS, unpublished). Major problems include the availability of sites, the effects of non-glacial factors on the sedimentary sequence, and limitations on the precision and resolution of the record.

The investigation of glacio-fluvial sequences has also been shown to have potential. MAIZELS & PETCH (1985) dated the inter-moraine zones in Austerdalen, South Norway, by lichenometry. Each area of glacio-fluvial sediments with their associated palaeochannel systems could be related to particular moraine ridges and hence to glacial history. THOMPSON & JONES (1986) demonstrated that glacio-fluvial terrace formation in the immediate proglacial area of Svinafellsjökull, in Iceland, was closely related to the frontal movements of the glacier. The most significant results have again been produced by A. Nesje and associates, who have identified several Neoglacial glacier expansion episodes from stratigraphic sequences in which glacio-fluvial sediments alternate with peat (NESJE et al., 1991; NESJE & DAHL, 1991a,b).

Yet another approach, based on the composition of moraines, has been suggested by MATTHEWS (1987a), who discovered that the roundness of clasts in "Little Ice Age" terminal moraines in Jotunheimen exhibits an altitudinal gradient and can be related to climate. As a first approximation, an increase in clast roundness of one unit is inferred to correspond to an increase in mean annual temperature of about 1.4°C within the study area. A mean roundness index of 2.0 (angular mode) for pyroxene-granulite clasts corresponds with a mean annual temperature of about -3.9°C, whereas a mean roundness index of 3.0 (subangular mode) represents a mean annual temperature of about -2.5°C. It is possible that this index could be developed for inferring palaeoclimates associated with the formation of older moraines.

3. Periglacial deposits

3.1 Solifluction lobe stratigraphy

The initiation of periglacial activity and changing rates of periglacial processes, both of which may be related to climatic variations, are receiving increasing attention in Northern Europe. Solifluction lobe stratigraphy has been the main source of information so far and has been the subject of detailed research on single lobes or a small number of lobes. Research has concentrated on the ^{14}C dating of soils progressively buried by the downslope advance of lobes. Initiation of solifluction lobe development has been variously dated to about 3000 ^{14}C yr B.P. (WORSLEY & HARRIS, 1974) and about 5000 ^{14}C yr B.P. (ELLIS, 1979a) in the Okstindan region of Northern Norway, about 3000 ^{14}C yr B.P. (NESJE et al., 1989) and later than about 1000 ^{14}C yr B.P. (MATTHEWS et al., 1986) in Southern Norway, and later than about 500 ^{14}C yr B.P. in the Fannich Mountains, Northwest Scotland (BALLANTYNE, 1986). Almost invariably the initiation of lobes has been attributed to climatic deterioration. Because of the relatively small number of dates, combined with problems in interpreting ^{14}C dates from soils, little can be said concerning changing rates of solifluction since lobe initiation.

There are, however, problems with interpreting the initiation of solifluction lobes in terms of climate. Unlike in the study of glacier variations, the physical basis of such interpretations is poorly understood. Indeed, the evidence for a climatic cause is often circumstantial and in the case of the Scottish evidence at least, grazing pressures may have been of crucial significance (cf. BALLANTYNE, 1991). Local site factors can be very influential and several processes may be involved. For example, in Leirdalen, Jotunheimen, initiation of the solifluction lobe dated by MATTHEWS et al. (1986) appears to have followed an interval of at least 2000 years during which colluvial deposits accumulated above a palaeosol that had probably developed throughout the Early Holocene. Gelifluction activity, which is more closely associated with a periglacial climate, was inferred to have occurred after 1000 ^{14}C yr B.P., but may have been accentuated by the close proximity of the glacier Storbreen.

NESJE et al. (1989) are of the opinion that the "Little Ice Age" was the most favourable period for gelifluction processes in the Jostedalsbreen region.

3.2 Other periglacial deposits

Other periglacial deposits have also been investigated for their palaeoclimatic implications. Two earth hummocks (thufur) have been ^{14}C dated in Breiseterdalen and Leirdalen, Jotunheimen (ELLIS, 1983) with the conclusion that deposition of hummock material and subsequent cryoturbation occurred some time after 4000 ^{14}C yr B.P. ELLIS suggests that although this conclusion is necessarily limited, future stratigraphic investigations of earth hummocks are capable of contributing to the emergence of a more precise pattern of Holocene environmental history in Scandinavia. A somewhat similar approach, with the addition of pollen analytical techniques, has provided data on the timing of palsa formation in Northern Norway (VORREN & VORREN, 1975; VORREN, 1979).

A recent attempt in the Veodalen area of Jotunheimen to date sorted patterned ground using lichenometry and weathering-based criteria (COOK, 1991) points to Early Holocene origins for these features. Nevertheless, results indicate at least one period of frost heaving and clast upheaval in circle centres which appears to have been associated with "Little Ice Age" climates. This study clearly suggests the possibility of using similar dating techniques on other sorted periglacial features with a suitably high boulder cover, e.g. rock glaciers and avalanche boulder tongues.

Rapid mass movement phenomena commonly associated with steep slopes in mountains may also yield evidence of periodic activity related to climate. Lichenometric dating has been used in several studies to date alpine debris flow activity in Northern Sweden (RAPP & NYBERG, 1961), Southern Norway (INNES, 1985) and Scotland (INNES, 1983a). RAPP & NYBERG (1961) identified at least four intervals of intensified activity within the last 2700 years. In Southern Norway, INNES (1985) found two periods of increased activity (1670-1720 and 1790-1860 A.D.). In both these areas a climatic cause seems possible, although there are dangers in assuming climatic changes where the underlying process depends on extreme rainfall events (ELLIS & RICHARDS, 1985). In Scotland, land use changes involving burning and overgrazing appear to explain the concentration of activity during the nineteenth and twentieth centuries (INNES, 1983a; see also BALLANTYNE, 1991). Dangers in utilizing only the surface record of such activity may be overcome if the opportunity arises to obtain stratigraphic evidence. Thus, in a study at the Storr on the Isle of Skye, Scotland, INNES (1983b) utilized ^{14}C dates on buried organic layers to identify three phases of "alluvial talus" accumulation. In this case, however, climatic change was considered an unlikely cause.

4. Soils

Their widespread occurrence, relatively slow decomposition rates, and the results of ^{14}C dating of palaeosols in subalpine and alpine environments in Southern Norway, all suggest considerable potential for the use of soils in palaeoenvironmental reconstruction near timberlines (MATTHEWS, 1985). ELLIS (1979b) recognizes five types of mountain soil in Norway: Podzols, Brown Soils, Humic Regosols, Regosols, and Bog Soils. Specific investigations of the potential of the first two of these soil types for both ^{14}C dating and pollen analysis have been made in Southern Norway.

In relation to Podzols (Humo-ferric Podzols), slow accumulation of the surface organic (FH) horizon resembles that of a peat and good pollen stratification provides an interpretable pollen record (CASELDINE, 1983; CASELDINE & MATTHEWS, 1985). Inferences can also be made about soil history based on ^{14}C dates from both FH and Bh (illuvial) horizons (ELLIS & MATTHEWS, 1984, Matthews, 1987b). At Haugabreen, investigations of similar Podzols beyond the Neoglacial ("Little Ice Age") glacier limit and near the present-day timberline, suggest that podzolisation began at about 5000 ^{14}C yr B.P. and spread to steeper slopes as late as 3500 ^{14}C yr B.P. as climate deteriorated (CASELDINE & MATTHEWS, 1987). Pollen analysis in the buried and surface profiles from these sites indicates a drop in the birch (*Betula pubescens*) tree limit from about 3600 to 3300 ^{14}C yr B.P. and after about 750 ^{14}C yr B.P. Between about 3300 and 750 ^{14}C yr B.P. the tree limit remained below or close to its present (660 m) altitude except for two possible upward extensions at about 2600 and ca. 2200-2000 ^{14}C yr B.P.

At Vestre Memurubreen, surface and buried alpine Brown Soils at 1480 m (300 m above the present birch tree limit) provide information on variations in the low-alpine/mid-alpine vegetation boundary (CASELDINE, 1984; MATTHEWS & CASELDINE, 1987) and indicate a possible climatic cooling of 2-4°C between ca. 5000 ^{14}C yr B.P. and the "Little Ice Age" maximum. The results also indicate that these Brown Soils accumulated slowly upwards, due to the addition of aeolian material, thus preserving a stratigraphic record.

Humic Regosols, which have yet to be analysed in this way in Scandinavia, may provide a similar record for areas dominated by snowbed communities. Regosols (thin organic crusts in areas of even later snow lie) have less potential, although one study has demonstrated that they can be used for ^{14}C dating and that they may contain an interpretable pollen assemblage (HARRIS et al., 1987).

5. Conclusion

An increasingly complex record of Holocene climatic variations is becoming available from physical proxy data from near timberline in Northern Europe. This is mainly based on the reconstruction of glacier variations, at present largely derived from the dating of moraine sequences (by historical and lichenometric methods) and moraine stratigraphic evidence (utilizing ^{14}C dates from palaeosols and plant remains). Increasing detail is being supplied from stratigraphic studies of glacio-lacustrine and glacio-fluvial sediments located downstream from glaciers, beyond Neoglacial moraine limits.

Regional differences in the pattern of Holocene glacier and climatic variations on a millennial timescale are emerging within the Scandinavian peninsula. These differences cannot be wholly explained in terms of different methods, approaches and interpretations. In Southern Norway and parts of maritime Northern Norway (especially Lyngen), the "Little Ice Age" glacier expansion exceeded earlier Neoglacial glacier expansions. In more continental locations of Northern Scandinavia (particularly in Northern Sweden), pre-"Little Ice Age" glacier expansion episodes were more extensive. These patterns appear to be explicable in terms of the geographical location of the glaciers in relation to latitudinal migration of the oceanic polar front, associated winter precipitation patterns, and continentality. It is suggested that Holocene glacier maxima tend to correspond with southerly migrations of the oceanic polar front. The importance of the "Little Ice Age" glacier expansion relative to earlier Neoglacial glacier expansion episodes in Southern Norway is accounted for by a more southerly position of the polar front at that time. Earlier glacier maxima in Northern Scandinavia imply less southerly positions of the polar front earlier in the Holocene.

Periglacial deposits have also been found of significance in palaeoenvironmental and palaeoclimatic reconstruction in Northern Europe. Several studies of solifluction lobe stratigraphy have indicated initiation of lobe formation and apparent climatic deterioration during the last 5000 years. Potentially useful deposits also include earth hummocks (thufur), sorted patterned ground and debris flow deposits. However, climatic triggers for the associated geomorphic processes are inadequately known.

Surface soils and buried palaeosols have been successfully used for ^{14}C dating, pollen analysis and palaeoenvironmental reconstruction (including changes in tree limits) in Southern Norway. Various soil types, including Podzols, Brown Soils and Humic Regosols, have considerable potential in these areas, not only for reconstructing vegetation history but also for reconstructing soil history, which may be affected by climatic change independently.

Acknowledgements

This paper is Jotunheimen Research Expeditions, Contribution No. 91. Of the many colleagues who have taken part in the research programme of the expeditions between 1970 and 1990, I would particularly like to thank Judith Cook and Danny McCarroll for the provision of unpublished data. Of those who work independently, I thank Colin Ballantyne (St. Andrews) and Atle Nesje (Bergen) for sight of unpublished manuscripts.

References

ALEXANDER, M. J. & WORSLEY, P. (1976): Glacier and environmental changes - Neoglacial data from the outermost moraine ridges at Engabreen, Northern Norway. Geogr. Ann. 58(A), 55-69

BALLANTYNE, C. K. (1986): Late Flandrian solifluction on the Fannich mountains, Ross-shire. Scott. J. Geol. 22, 395-406

BALLANTYNE, C. K. (1990): The Holocene glacial history of Lyngshalvöya, Northern Norway: chronology and climatic implications. Boreas 19, 93-117

BALLANTYNE, C. K. (1991): Late Holocene erosion in upland Britain: climatic deterioration or human influence? The Holocene 1, 81-85

BØGEN, J.; WOLD, B. & ØSTREM, G. (1989): Historical glacier variations in Scandinavia. In: Oerlemans, J. (ed.): Glacier fluctuations and climatic change, Kluwer, Dordrecht, 109-128

CASELDINE, C. J. (1983): Pollen analysis and rates of pollen incorporation into a radiocarbon-dated palaeopodzolic soil at Haugabreen, Southern Norway. Boreas 12, 233-246

CASELDINE, C. J. (1984): Pollen analysis of a buried arctic-alpine brown soil from Vestre Memurubreen, Jotunheimen, Norway: evidence for Postglacial high-altitude vegetation change. Arct. Alp. Res. 16, 423-430

CASELDINE, C. J. & MATTHEWS, J. A. (1985): ^{14}C dating of palaeosols, pollen analysis and landscape change: studies from the low- and mid-alpine belts of Southern Norway. In: Boardman, J. (ed.) : Soils and Quaternary landscape evolution, Wiley, Chichester, 87-116

CASELDINE, C. J. & MATTHEWS, J. A. (1987): Podzol development, vegetation change and glacier variations at Haugabreen, Southern Norway. Boreas 16, 215-230

COOK, J. D. (1991): Sorted circles, relative-age dating and palaeoenvironmental reconstruction in a periglacial alpine environment, Eastern Jotunheimen, Norway: lichenometric and weathering-based approaches. The Holocene 1, 128-141

ELLIS, S. (1979a): Radiocarbon dating evidence for the initiation of solifluction ca. 5500 years B.P. at Okstindan, North Norway. Geogr. Ann. 61 A, 29-33

ELLIS, S. (1979b): The identification of some Norwegian mountain soil types. Norsk Geogr. Tidsskr. 33, 205-212

ELLIS, S. (1983): Stratigraphy and ^{14}C dating of two earth hummocks, Jotunheimen, South Central Norway. Geogr. Ann. 65 A, 279-287

ELLIS, S. & MATTHEWS, J. A. (1984): Pedogenic implications of a ^{14}C dated palaeopodzolic soil at Haugabreen, Southern Norway. Arct. Alp. Res. 16, 77-91

ELLIS, S. & RICHARDS, K. S. (1985): Pedogenic and geotechnical aspects of Late Flandrian slope instability in Ulvådalen, West-Central Norway. In: Richards, K. S.; Arnett, R. R. & Ellis, S. (eds.): Geomorphology and soils, George Allen & Unwin, London, 328-347

ERIKSTAD, L. & SOLLID, J. L. (1986): Neoglaciation in South Norway using lichenometric methods. Norsk Geogr. Tidsskr. 40, 85-105

GRIFFEY, N. J. (1976): Stratigraphical evidence for an early Neoglacial glacier maximum of Steikvassbreen, Okstindan, Northern Norway. Norsk Geol. Tidsskr. 56, 187-194

GRIFFEY, N. J. & MATTHEWS, J. A. (1978): Major Neoglacial glacier expansion episodes in Southern Norway: evidences from moraine ridge stratigraphy with ^{14}C dates on buried palaeosols and moss layers. Geogr. Ann. 60 A, 73-90

GRIFFEY, N. J. & WORSLEY, P. (1978): The pattern of Neoglacial glacier variations in the Okstindan region of Northern Norway during the last three millennia. Boreas 7, 1-17

GROVE, J. M. (1985): The timing of the Little Ice Age in Scandinavia. In: Tooley, M. J. & Sheail, G. M. (eds): The climatic scene. George Allen & Unwin, London, 132-153

HARRIS, C.; CASELDINE, C. J. & CHAMBERS, W. (1987): Radiocarbon dating of a palaeosol buried by sediments of a former ice-dammed lake, Leirbreen, Southern Norway. Norsk Geogr. Tidsskr. 41, 81-90

INNES, J. L. (1983a): Lichenometric dating of debris-flow deposits in the Scottish Highlands. Earth Surface Proc. Landf. 8, 579-588

INNES, J. L. (1983b): Stratigraphic evidence of episodic talus accumulation on the Isle of Skye, Scotland. Earth Surface Proc. Landf. 8, 399-403

INNES, J. L. (1985): Lichenometric dating of debris-flow deposits in alpine colluvial fans in Southwest Norway. Earth Surface Proc. Landf. 10, 519-524

INNES, J. L. (1984): Lichenometric dating of moraine ridges in Northern Norway: some problems of application. Geogr. Ann. 66 A, 341-352

KARLÉN, W. (1973): Holocene glacier and climatic variations, Kebnekaise mountains, Swedish Lappland. Geogr. Ann. 55 A, 29-63

KARLÉN, W. (1975): Lichenometrisk datering i norra Skandinavien - metodens tillförlitlighet och regionala tillåmpning. Department of Physical Geography, Univ. of Stockholm, Report No. 22

KARLÉN, W. (1976): Lacustrine sediments and tree-limit variations as indicators of Holocene climatic fluctuations in Lappland: Northern Sweden. Geogr. Ann. 58 A, 1-34

KARLÉN, W. (1979): Glacier variations in the Svartisen area, Northern Norway. Geogr. Ann. 61 A, 11-28

KARLÉN, W. (1981): Lacustrine sediment studies. Geogr. Ann. 63 A, 273-281

KARLÉN, W. (1982): Holocene glacier fluctuations in Scandinavia. Striae 18, 26-34

KARLÉN, W. (1988): Scandinavian glacial and climatic fluctuations during the Holocene. Quat. Sci. Rev. 7, 199-209

KARLÉN, W. & DENTON, G. H. (1975): Holocene glacial variations in Sarek National Park, Northern Sweden. Boreas 5, 25-56

LAMB, H. H. (1979): Climatic variation and changes in the wind and ocean circulation: the Little Ice Age in the Northerneast Atlantic. Quat. Res. 11, 1-20

MAIZELS, J. K. & PETCH, J. R. (1985): Age determination of intermoraine areas, Austerdalen, Southern Norway. Boreas 14, 51-65

MATTHEWS, J. A. (1980): Some problems and implications of ^{14}C dates from a podzol buried beneath an end moraine at Haugabreen, Southern Norway. Geogr. Ann. 62 A, 185-208

MATTHEWS, J. A. (1982): Soil dating and glacier variations: a reply to Wibjörn Karlén. Geogr. Ann. 64 A, 15-20

MATTHEWS, J. A. (1984): Limitations of ^{14}C dates from buried soils in reconstructing glacier variations and Holocene climate. In: Mörner, N. A. & Karlén, W. (eds): Climatic change on a yearly to millennial basis. Reidel, Dordrecht, 281-190

MATTHEWS, J. A. (1985): Radiocarbon dating of surface and buried soils: principles, problems and prospects. In: Richards, K. S., Arnett, R. R. & Ellis, S. (eds): Geomorphology and soils. George Allen & Unwin, London, 269-288

MATTHEWS, J. A. (1987a): Regional variation in the composition of Neoglacial end moraines, Jotunheimen, Norway: an altitudinal gradient in clast roundness and its possible palaeoclimatic significance. Boreas 16, 173-188

MATTHEWS, J. A. (1987b): Some aspects of the ^{14}C dating of buried soil horizons. In: Stevens, P. (ed.): Soils and the time factor. (Inst. of Terrestrial Ecology, Bangor, U.K.), 13-30

MATTHEWS, J. A. (1989): Holocene glacier variations in Southern Norway: ^{14}C dating of the Late Neoglacial glacier maximum. Terra Abstracts 1/1, 426

MATTHEWS, J. A. (1990): Holocene glacier and climatic variations in Southern Scandinavia: review and new evidence from moraine stratigraphy. 3rd Nordic conference on climatic change and related problems, Univ. of Tromsø, Norway, April 2-4, 1990. Programme and Abstracts, 22

MATTHEWS, J. A. (1991): The Late Neoglacial ("Little Ice Age") glacier maximum in Southern Norway: new ^{14}C-dating evidence and climatic implications. The Holocene 1, 219-233

MATTHEWS, J. A. & CASELDINE, C. J. (1987): Arctic-alpine Brown Soils as sources of palaeoenvironmental information: further ^{14}C dating and palynological evidence from Vestre Memurubreen, Jotunheimen, Norway. J. Quat. Sci. 2, 59-71

MATTHEWS, J. A. & DRESSER, P. Q. (1983): Intensive ^{14}C dating of a buried palaeosol horizon. Geol. Fören. Stockh. Förh. 105, 59-63

MATTHEWS, J. A.; HARRIS, C. & BALLANTYNE, C. K. (1986): Studies on a gelifluction lobe, Jotunheimen, Norway: ^{14}C chronology, stratigraphy, sedimentology and palaeoenvironment. Geogr. Ann. 68 A, 345-360

MATTHEWS, J. A. & SHAKESBY, R. A. (1984): The status of the "Little Ice Age" in Southern Norway: relative-age dating of Neoglacial moraines with Schmidt hammer and lichenometry. Boreas 13, 333-346

MCCARROLL, D. (1989): Potential and limitations of the Schmidt hammer for relative-age dating: field tests on Neoglacial moraines, Jotunheimen, Southern Norway. Arct. Alp. Res. 21, 268-275

MCCARROLL, D. (1991): The age and origin of Neoglacial moraines in Jotunheimen, Southern Norway: new evidence from weathering-rind data. Boreas 20, 283-295

NESJE, A. & DAHL, S. O. (1991a): A record of Holocene glacier variations of Blåisen, Hardangerjøkulen, Central Southern Norway: evidence for mid-Holocene glacier contraction. Quat. Res. 35, 25-40

NESJE, A. & DAHL, S. O. (1991b): Late Holocene glacier fluctuations in Bevringsdalen, Jostedalsbreen region, Western Norway (ca. 3200-1400 yr. B.P.). The Holocene 1, 1-7

NESJE, A.; KVAMME, M. & RYE, N. (1989): Neoglacial gelifluction in the Jostedalsbreen region, Western Norway: evidence from dated buried palaeopodsols. Earth Surface Proc. Landf. 14, 259-270

NESJE, A.; RYE, N.; KVAMME, M. & LØVLIE, R. (1991): Holocene glacial and climate history of the Jostedalsbreen region, western Norway: evidence from lake sediments and terrestrial deposits. Quat. Sci. Rev. 10, 87-114

ØSTREM, G. (1965): Problems of dating ice-cored moraines. Geogr. Ann. 67 A, 1-38

RAPP, A. & NYBERG, R. (1961): Alpine debris flows in Northern Scandinavia: morphology and dating by lichenometry. Geogr. Ann. 63 A, 183-196

THOMPSON, A. & JONES, A. (1986): Rates and causes of proglacial river terrace formation in Southeast Iceland: an application of lichenometric dating techniques. Boreas 15, 231-246

VORREN, K.-D. (1979): Recent palsa datings, a brief survey. Norsk Geogr. Tidsskr. 33, 217-219

VORREN, K.-D. & VORREN, B. (1975): The problem of dating a palsa. Two attempts involving pollen diagrams, determination of moss subfossils, and ^{14}C dating. Astarte 8, 73-81

WORSLEY, P. & ALEXANDER, M. J. (1976): Neoglacial palaeo- environmental change at Engabrevatn, Svartisen Holandsfjord, North Norway. Norsk Geol. Unders. 321, 37-66

WORSLEY, P. & HARRIS, C. (1974): Evidence for Neoglacial solifluction at Okstindan, North Norway. Arctic 27, 128-144

Address of the author:

Dr. J. A. Matthews, Department of Geology, Univ. of Wales College of Cardiff (UWCC), P.O. Box 914, Cardiff CF1 3YE, Wales, U.K.

Holocene forest limit fluctuations and glacier development in the mountains of Southern Norway, and their relevance to climate history

Mons Kvamme

Summary

Recent papers concerning Holocene forest history and glacier and climatic development in the mountains of South Norway are reviewed. All investigations conclude that some Holocene climatic optimum has occurred when the upper pine-forest limit was situated 150 to 300 m higher than the present elevation in oceanic and continental areas, respectively. Regional summer temperature is estimated to have been 1.5-2.0°C higher than today. This most favourable climatic period in the Holocene is situated in the Late Boreal or Early Atlantic chronozone. A curve of Holocene equilibrium-line fluctuations at the Jostedalsbreen is presented, indicating that this ice cap had melted away at that time. The date of climatic deterioration is discussed. According to most authors this started in the mid-Holocene, possibly already prior to 5000 yr B.P. The development in South Norway is compared with results from Central Sweden and the Alps. Based on available data, the existence of a belt of subalpine birch forest through the entire Holocene is discussed and questioned.

Zusammenfassung

In diesem Beitrag werden die neuesten Untersuchungen über die holozäne Waldgeschichte und die Gletscher- und Klimaentwicklung in den Gebirgen Südnorwegens kritisch besprochen. Alle Untersuchungen deuten klar auf ein holozänes Klimaoptimum hin, in dessen Verlauf sich die obere Kiefernwaldgrenze sowohl in den ozeanischen als auch in den kontinentalen Regionen um 150-300 m über der heutigen befand. Die Sommertemperaturen dürften in der untersuchten Region die heutigen Temperaturen um ungefähr 1,5-2,0°C überschritten haben. Diese klimatisch günstigste Phase des Holozän lag im Spätboreal oder frühen Atlantikum. Es wird eine Kurve der holozänen Schwankungen der Gleichgewichtslinie zwischen Ablation und Akkumulation des Jostedalsbreen-Gletschers vorgestellt, die zeigt, daß diese Eiskappe damals abgeschmolzen war. Der Zeitpunkt der Klimaverschlechterung wird diskutiert. Nach Meinung der meisten Autoren begann diese im mittleren Holozän, vielleicht bereits vor 5000 J.v.h. Die Entwicklung in Südnorwegen wird mit Ergebnissen aus Zentralschweden und den Alpen verglichen. Auf der Basis des vorhandenen Materials wird die Existenz einer subalpinen Birkenstufe durch das gesamte Holozän hindurch diskutiert und hinterfragt.

1. Introduction

The mountains of Southern Norway are relatively low with only a limited number of the summits exceeding 2000 m a.s.l. Due to the northern latitudes, however, large areas are unforested because of climatic factors. The uppermost limits of the subalpine birch forests in all the Scandinavian mountains are situated at an altitude not higher than 1220 m a.s.l. in the central part of Southern Norway (VE, 1930; AAS, 1964; ODLAND et al., in press). From here, there is a strong gradient to the coast of Western Norway, where the forest limits lie below 500 m a.s.l. (AAS & FAARLUND, 1988). Toward the east, the upper subalpine birch forest limit decreases less rapidly to an elevation of about 900 m a.s.l. (KULLMAN, 1989; AAS & FAARLUND, 1988).

The situation has not always been like this. Occurrences of high-altitude pine-stumps, sometimes at elevations higher than the present subalpine birch forest, are an obvious demonstration of the effect of an earlier, warmer climate on the vegetation. The age and elevation of the mountain forest belt fluctuations through the Holocene have been much discussed, as have the vegetational composition at the forest/alpine tundra ecotone (e.g. NORDHAGEN, 1943; FÆGRI, 1945; HAFSTEN, 1960; 1965; MOE, 1973; 1978, 1979; INDRELID & MOE, 1982; MOE & ODLAND, in press.; SIMONSEN, 1980; SELSING & WISHMAN, 1984; SELSING, 1986; AAS & FAARLUND, 1988). In this paper the main publications from the last 15 years concerning these questions are discussed and compared with results from areas at the Jostedalsbreen in Western Norway and some other mountain areas in Europe.

2. Pollen and pine-stump data from the mountain chain of Southern Norway

Earlier pollen-analytical investigations documented the existence of pine and grey alder (*Alnus incana*) forests at higher altitudes than today (FÆGRI, 1945; HAFSTEN, 1965). SIMONSEN (1980) and MOE (1978, 1979) provided the first extensive studies on the Holocene mountain forest history in Southern Norway. The work of SIMONSEN was based on pollen diagrams from different altitudes in the Hardangerfjord area (Fig. 1). Because he only had four radiocarbon dates, he had to rely heavily upon relative pollen stratigraphy in his conclusions. His main results concerning the high-altitude forest history were:

a) *Betula* dominated at the forest limit approximately 900 m a.s.l. in the earliest Holocene;
b) Pine forest, mixed with some birch and *Ulmus* reached 900 to 1000 m a.s.l. in the Early- and mid-Atlantic chronozone (chronozones according to MANGERUD et al., 1974), and this combination comprised the forest composition of the forest/alpine-tundra ecotone at that time;
c) the upper limit of pine forest decreased between 200 and 300 m in the Late Atlantic and Subboreal chronozones, with birch forest developing in a zone with approximately 200 m vertical extension above the pine forest (Fig. 2).

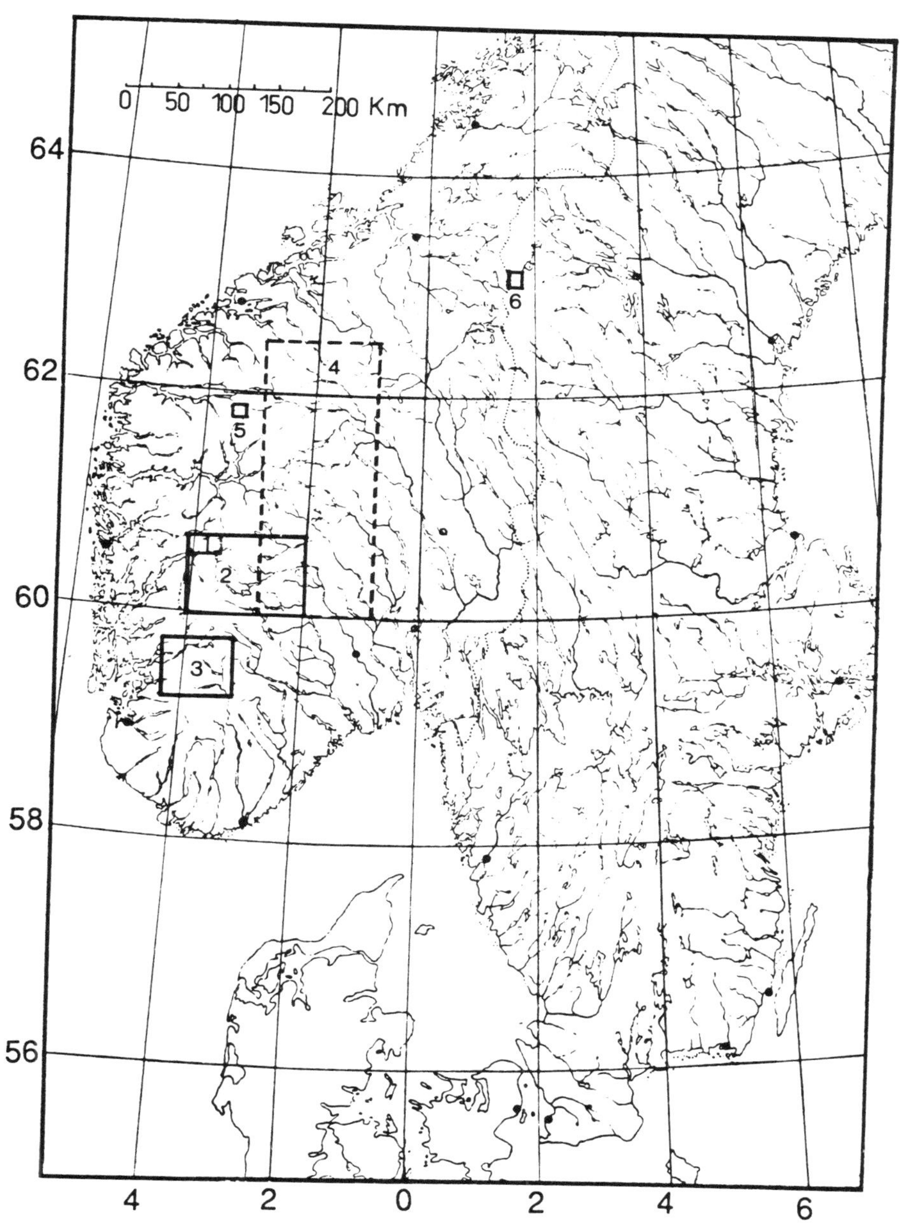

Fig. 1 Investigated areas. (1) SIMONSEN, 1980; (2) MOE, 1978; (3) SELSING & WISHMAN, 1984; (4) AAS & FAARLUND, 1988; (5) NESJE & KVAMME, in press; (6) KULLMAN, 1988,1989,1990)

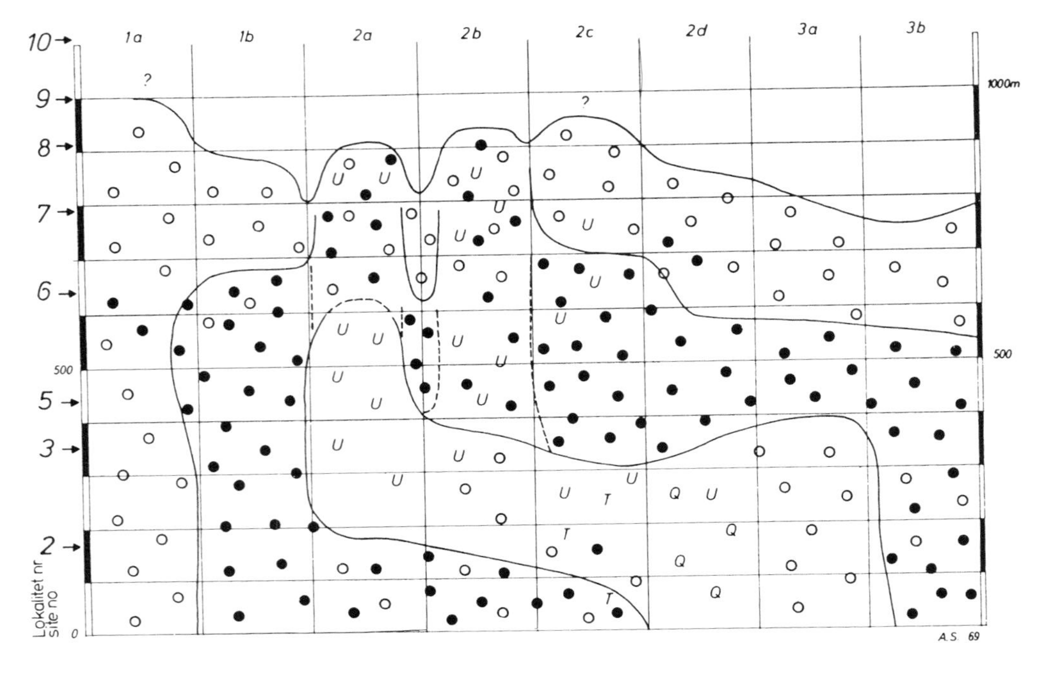

○ Bjørk (*Betula*) ● Furu (*Pinus*) *U* Alm (*Ulmus*) *T* Lind (*Tilia*) *Q* Eik (*Quercus*)

Fig. 2 Holocene development of forest limit fluctuations and composition in the Hardangerfjord area. From SIMONSEN, 1980

MOE worked on the Hardangervidda plateau 1100-1250 m a.s.l., an area that today is mostly treeless. Based on pollen diagrams between 900 and 1300 m a.s.l., pine stumps, and radiocarbon datings, he concluded that before 8000 yr B.P. pine forest expanded to about 1250 m a.s.l., which is about 250 m higher than its present upper limit. During the Atlantic chronozone, pine forest descended slightly and became a mixture with some deciduous trees. During a short period before 5000 yr B.P. there are indications of a new, minor pine expansion to higher altitudes. However, MOE found the same decrease in pine forest limit after 5000 yr B.P. as SIMONSEN had reported, accompanied by the development of an upper subalpine birch forest.

In a recent paper (MOE & ODLAND, in press), a summary is made of all relevant data from FÆGRI (1945), HAFSTEN (1965), SIMONSEN (1980), MOE (1973, 1978), and INDRELID & MOE (1982). Two new curves are presented (Fig. 3) of the mountain forest composition and the Holocene forest/alpine-tundra ecotone altitudinal fluctuations in the Hardangervidda area.

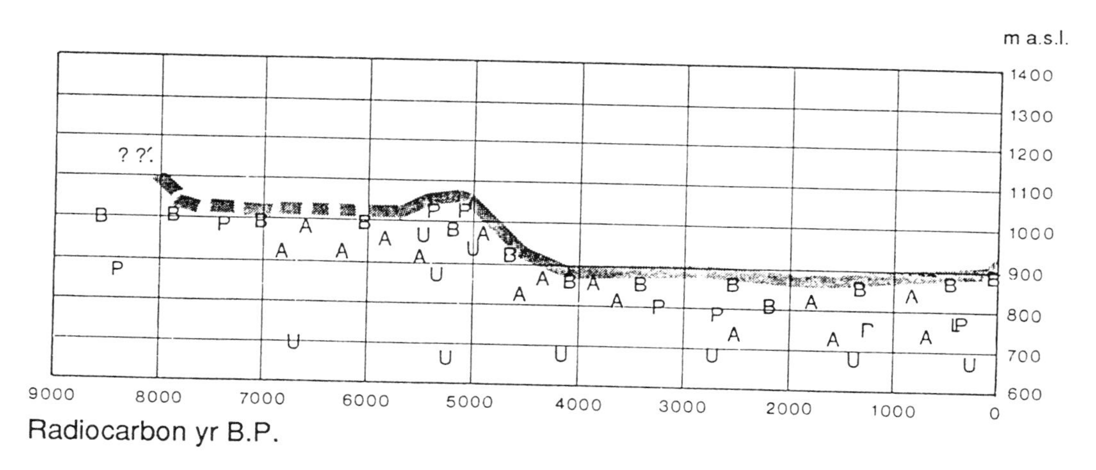

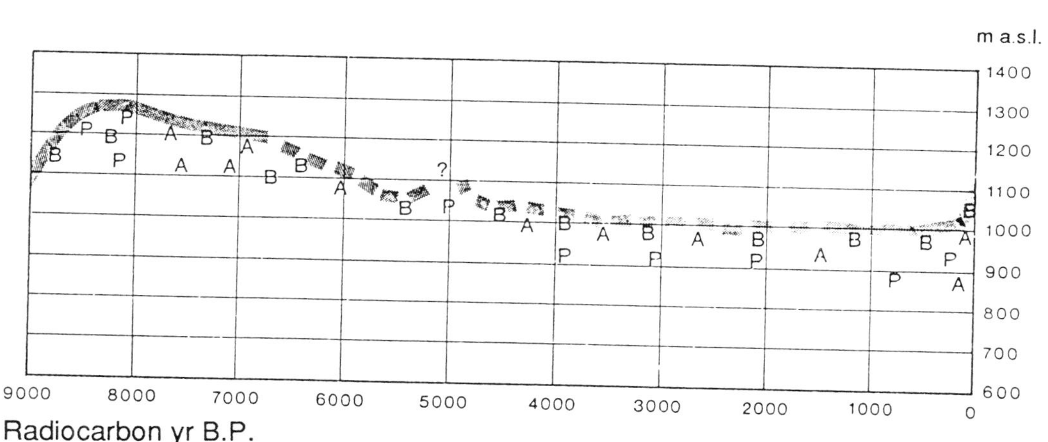

Fig. 3 Forest limit fluctuations on the western (top) and eastern (below) part of Hardangervidda (MOE & ODLAND, in press). B=*Betula*, P=*Pinus*, A=*Alnus*, U=*Ulmus*

In addition, the relation between *Alnus incana* and summer temperature has been studied. The results show a strong positive correlation between the upper limit of *Alnus incana* forest today and the mean maximum temperature of July at the nearest lowland meteorological station. At the forest limit however, this temperature appears to be constant, calculated to 17,2±0.8°C independent of altitude (MOE & ODLAND, in press.). Based on all available results from the area, they have calculated the difference in summer temperature between the Atlantic chronozone and today to be 1,7°C, corrected for Postglacial isostatic rebound.

From areas southwest of Hardangervidda, SELSING (1986) has published a series of pine-stump datings. The highest stumps are dated to the Early Atlantic chronozone and all younger datings come from lower altitudes. These results confirm the general trend from Hardangervidda (Fig. 4). Furthermore, SELSING & WISHMAN (1984) have shown that between these areas there is a strong climatic gradient today, with the upper limit of the present pine-forest situated at about 600 m a.s.l. to the west and 930 m a.s.l to the east. Based on the geographical distribution of pine-stumps, they found that during the Atlantic chronozone, the pine forest was situated about 100 m higher in the west and 150 m higher in the eastern part of their study area. This implies that the west-east climatic gradient was stronger during the Atlantic chronozone than today. They also concluded that with such a strong west-east gradient, the summer temperature/altitude gradient must have been steeper than today in the eastern areas during the Atlantic. Thus the difference in summer temperature between today and the Atlantic chronozone was probably greater in lowland situations than at higher altitudes.

AAS & FAARLUND (1988) have published a large number of radiocarbon-dated samples from high-altitude subfossil pine and birch remains (Fig. 4). These have been collected from the eastern part of the mountain ranges in Southern Norway, an area where the most continental climate in all of Norway is found. Based on the altitudinal distribution of stumps, they concluded that pine forests during the Atlantic chronozone were growing 300 m higher than today in these areas. They calculated the summer temperature to have been 1.4-1.8°C higher than today, corrected for Postglacial isostatic rebound. They also proposed that above the pine forest, subalpine birch forests had prevailed through the entire Holocene. I discuss this proposal later.

3. Investigations around the Jostedalsbreen glacier, Western Norway

In the vicinity of large glaciers, like Jostedalsbreen, the presence of the ice will influence local climate and hence local vegetation. Thus, the limits of birch and pine forests in the Jostedalen valley are depressed 120 and 150 m, respectively, over a distance of 25 km close to the glacier (ODLAND et al., 1989). Because of the strong impact of local climate on the vegetation, studies of local vegetation history have proved particularly useful, in conjunction with Quaternary geological investigations, in reconstructing the Holocene glacial development on the Jostedalsbreen plateau and thus the regional climate history (KVAMME, 1984, 1989, in prep.; NESJE et al., 1991; NESJE & KVAMME, in press). The main conclusions of these investigations are given below.

At the northeastern end of the glacier, in Stryn, a small, infilled pond 700 m a.s.l. has been investigated. It lies 2 km from and 260 m below the frontal position of the glacier during the Little Ice Age. Today the locality is situated close to the upper limit of the subalpine birch forest, but the pollen diagram shows that through the later part of the Atlantic chronozone, *Ulmus* was growing locally at or near the site (KVAMME, 1984). Even when corrected for Postglacial isostatic rebound, such occurrences suggest a summer temperature 3.0-3.5 (3.2)°C warmer than today. This estimate includes the present effect of the glacier on local climate. Without this effect, the temperature difference can be estimated to be 1.5-2.0 (1.8)°C. Applying a summer temperature/altitude lapse-rate of 0.65 °C/100 m^{-1} (close to the mean of the two values given by SELSING & WISHMAN, 1984), gives an Atlantic chronozone summer temperature at the present ELA (equilibrium line altitude) of 2.7°C more than today, measured by NVE on the glacier (ROALD, 1973). This implies an Atlantic chronozone ELA approximately 400 m higher than today, i.e. above 1960 m a.s.l. As only a few summits exceed this altitude, most of the Jostedalen ice cap probably melted away during the Atlantic chronozone (KVAMME, 1984; NESJE et al., 1991; NESJE & KVAMME, in press; KVAMME, in prep).

From the same locality there is evidence for sediment contamination, resulting in radiocarbon dates from the upper parts of the sediment that are too old. This probably started shortly before 5000 yr B.P. From this time old carbonaceous material appears to have drained into the pond. This material also contained pollen, and the curves above this level make little sense (KVAMME, 1984). As most other explanations (e.g. anthropogenic erosion) can certainly be excluded, this contamination probably resulted from climatic change. Processes associated with frost action and snow line are active at the locality today (KVAMME, 1984, in prep.).

From the Jostedalen valley, situated across the glacier south of the Stryn locality, pollen diagrams have been prepared from four sites between 500 m a.s.l. and 800 m a.s.l. Although they are all very local, there is a common and obvious trend in them all prior to 5000 yr B.P. *Alnus incana*, previously a co-dominant tree at these altitudes, declined in its importance and, to a large extent, disappeared from the local vegetation between 6000 yr B.P. and 5000 yr B.P. A few stands persist at favourable places, but it is no longer important as a forest tree in the valley above 500 m a.s.l. (ODLAND et al., 1989). This appears to have been the situation for the last 5000 years, and in the same period *Betula pubescens* has dominated the forests at these altitudes (KVAMME, 1989, in prep.).

The altitudinal range of *Alnus incana* forest thus decreased at least 200 m during less than 1000 years prior to 5000 yr B.P. During the same time, climatic change caused the onset of sediment contamination on the other side of the glacier. In conjunction with other data from the area (NESJE et al., 1991; CASELDINE & MATTHEWS, 1987), these observations suggest that strong and relatively rapid environmental changes took place between 6000 and 5000 yr B.P. in the areas today bordering the eastern part of the Jostedalsbreen. A general climatic deterioration probably started about 6000 yr B.P., and after some time the critical glaciation threshold for the Jostedalsbreen plateau was passed. This initiated the formation of the present Jostedalsbreen glacier, which generated an additional, local contribution to the regional effect of the general climatic decline. Therefore, the observed environmental changes, in for example the vegetation, are somewhat more pronounced here than in many other areas during the same period. Based on all the available data (NESJE et al., 1991; NESJE & KVAMME, in press; KVAMME, in prep.), a curve for the Holocene ELA fluctuations of the eastern part of the Jostedalbreen has been constructed (Fig. 5).

4. Discussion

The forest/alpine tundra ecotone is a very dynamic biological boundary, influenced by many factors in addition to climate, as soil, topography, snow conditions, and human impact. The uppermost limit of solitary trees is a probably better indicator of regional climate (KULLMAN, 1979, 1990). With "proxy data" from pollen analysis and plant-macrofossil studies, it is usually not possible to detect solitary trees. Most other factors influencing the forest limit normally can be detected when working on a local scale and hence are excluded from the calculations (KVAMME, 1989, in prep.). Less uncertainty can be attached to climatic reconstructions when they are based on data collected from a limited area rather than from a wide and varied geographical region.

The investigations in Southern Norway have been made at very different geographical scales, but their results have some important features in common. First, they all indicate that there has been a period with significantly higher tree and forest limits than today. This is primarily explained by higher summer temperatures. Second, this "climatic optimum" is placed in the Early Holocene, in the Late Boreal or Early Atlantic chronozones. Third, most estimates indicate a summer temperature 1.5-2.0°C higher than today. SELSING & WISHMAN (1984) have lower estimates, probably because they worked in areas with strong gradients in oceanicity.

This occurrence of a climatic optimum in the South Norwegian mountains in the Early Holocene contrasts with earlier opinions (e.g. HAFSTEN, 1960, 1965), but is in good accordance with modern independent data from general circulation simulation models of climate based on geophysical and atmospheric research (KUTZBACH & GUETTER, 1986; KUTZBACH & GALLIMORE, 1988; COHMAP, 1988). These models also indicate a higher seasonality in the first part of Holocene (i.e. prior to 6000 yr B.P.). From Central Sweden KULLMAN (1988, 1989, 1990) has shown a pattern of Holocene climatic development that closely resembles that for Southern Norway. In particular, there is a good correspondence between his curve of Holocene tree limit fluctuations (KULLMAN, 1990: 105) and the curve of Holocene ELA fluctuations at Jostedalsbreen (Fig. 5). Of the investigations discussed in this paper, these are the two performed at the finest geographical scale.

In recent papers BIRKS (1986, 1988, 1989, 1990) has demonstrated the "complex, unpredictable, and highly individualistic responses of different trees to the major climatic changes of the Early Holocene" (BIRKS, 1990: 145). In his 1990 paper he suggested that

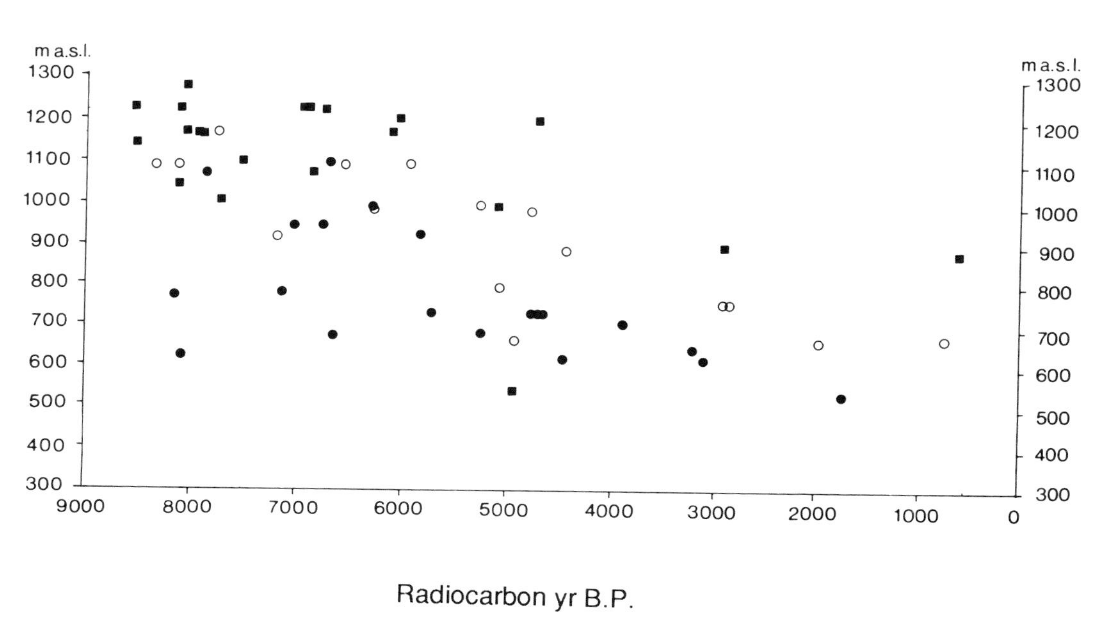

Fig. 4 Radiocarbon dated subfossil samples of pine (*Pinus*) from Southern Norway, published by MOE (1978) ○, SELSING (1986) ●, and AAS & FAARLUND (1988) ■

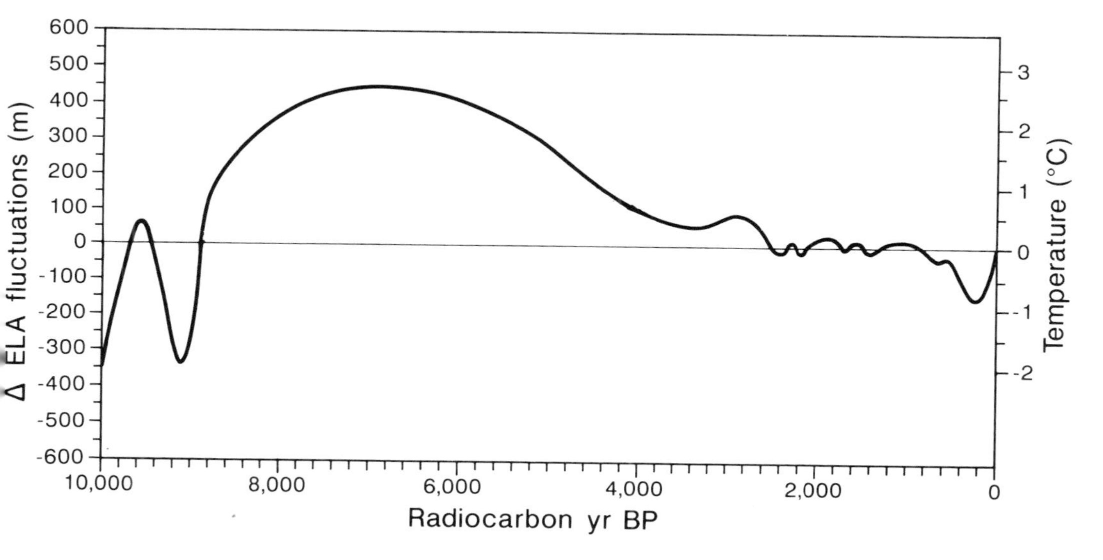

Fig. 5 Holocene ELA fluctuations at the eastern part of Jostedalsbreen (NESJE & KVAMME, in press)

there were hardly any climatic limitations for the altitudinal expansion of pine at the time of its arrival in different parts of Scotland and Fennoscandia. In the same paper he showed that the altitudinal distribution of published radiocarbon-dated pine-stumps from South Norway corresponds very well with theoretical expectations, using the climatic simulations of KUTZBACH & GUETTER (1986) for estimating the Scandinavian climate for 9000, 6000, and 3000 yr B.P. The explanation may be that the best period for the altitudinal expansion of pine fell between the simulation data of 9000 and 6000 yr B.P. At this time, probably around 8000 yr B.P., the distribution of pine was more a question of its ability to take advantage of the available conditions than of any climatic constraints. The time lag between climatic improvement and arrival of tree species can be important in this connection (BIRKS, 1989, 1990). This has also been shown at the Jostedalsbreen (KVAMME, 1989, in prep.).

According to earlier views, the main Holocene climatic deterioration occurred rather late, about at the time of the transition to the Subatlantic chronozone (e.g. HAFSTEN, 1960). More recent investigations have concluded that this deterioration took place at an earlier date, the age of which is still being discussed. MOE (1978, 1979), and MOE & ODLAND (in press.), propose that growing conditions for pine on Hardangervidda had deteriorated already at the beginning of the Atlantic chronozone, but the main climatic deterioration took place shortly after 5000 yr B.P. AAS & FAARLUND (1988) suggest a decrease in summer temperature at an even later date, although their documentation for this appears rather scanty (8 of 37 radiocarbon datings are younger than 5000 yr B.P. In addition they rely on the tentative curve of ELA variations of LIESTØL, 1960, 1969)).

From more oceanic areas the pattern is somewhat different. In Central Sweden (with greater oceanicity than in Jotunheimen/Hardangervidda), KULLMAN (1990) has shown that tree lines started to decline about 6000 yr B.P. He interprets this as a response to lower summer temperatures and reduced seasonality. The ELA calculations at Jostedalsbreen (Fig. 5) also point to a climatic deterioration starting about 6000 yr B.P. Here it can be shown that it primarily must have been a result of reduced summer temperatures, in combination with longer snow cover and probably increased humidity (NESJE & KVAMME, in press.; KVAMME, in prep.). As the earliest indications of climatic deterioration come from the oceanic areas, it may be possible that the mid-Holocene climatic change involved a change to a higher humidity in addition to lower summer temperatures. This contradicts the earlier view of the Subboreal chronozone as a dry and warm period.

The pattern of Scandinavian Holocene climatic development outlined above differs substantially from the proposed development in the Tyrolian Alps published by PATZELT & BORTENSCHLAGER (1973), PATZELT (1974), and BORTENSCHLAGER (1982, 1989). There they have no climatic optimum, but postulate that the tree limits, snow-lines, and summer temperatures have been fluctuating about the values of today throughout the Holocene. From the Swiss Central Alps BURGA (1988) argues that there has been a moderate mid-

Holocene (6000-4800 yr B.P.) climatic optimum, followed by a decrease in the potential timberline of 100-200 m. On the other hand, from different data sources PORTER & OROMBELLI (1985) and HUNTLEY & PRENTICE (1988) have calculated a July temperature of the Atlantic chronozone 4-5°C warmer than today for the Western Italian Alps and the Alps in general, respectively. However, the pattern of Holocene climatic development in the Alps does not necessarily have to be the same as in the Scandinavian mountains. In Scandinavia, the climatic conditions of the mountains, and in particular of the glaciers, are primarily influenced by two factors: (1) the position of the Polar Front and (2) the cyclone tracks of the North Atlantic circulation system. It is unlikely that either of these have had any influence on the climate and glaciers of the Alps.

5. The history of the subalpine birch forest

AAS & FAARLUND (1988) argue that a belt of subalpine birch forest existed in the Scandinavian mountains since the beginning of the Holocene. They base this suggestion on the presence of birch wood in Jotunheimen at altitudes 200 m above the limit of subalpine birch forest of today, and the Preboreal chronozone dominance of *Betula* in nearly all Scandinavian pollen diagrams. The composition of the Preboreal vegetation is beyond the scope of this paper. However, conditions at that time (IVERSEN, 1960, 1973; DANIELSEN, 1970; BIRKS, 1986; PAUS, 1988) were so different from the situation in the subalpine birch forest of today that any comparison is of very limited value.

The high-altitude subfossil birch remains are, on the other hand, important evidence for the existence of birch trees at these altitudes in the Late Boreal and Early Atlantic chronozones. The interpretation of these finds as birch forest may be somewhat bold, in particular as they are scattered over a greater part of the East Norwegian mountain range. If a subalpine birch belt had existed through the Holocene, it is reasonable to expect that in the same areas birch macrofossils should be found at higher altitudes than pine of similar age. In the data presented by AAS & FAARLUND (1988) only three areas had birch and pine of the same age. In two of these they were from a similar altitude, in the third area *Betula* was found 12 m above the *Pinus* remains.

In addition they refer to BARTH et al. (1980) who have published two wood samples from Rondane, one of birch (1230 m a.s.l.) and one of pine (1040 m a.s.l.), both radiocarbon dated to the Late Boreal chronozone and situated far above their present respective forest limits. However, pollen analysis of the peat containing the birch sample, contained 42% *Pinus* and only 10% *Betula* pollen (BARTH et al., 1980). Both SIMONSEN (1980) and KVAMME (1988, 1989) have shown that at localities within the subalpine birch forest, the pollen deposition is totally dominated by *Betula*.

Several pollen diagrams published by MOE (1978) from Hardangervidda are situated between the upper limit of the present subalpine birch forest (1000-1100 m a.s.l.) and the highermost level of the Early Holocene pine forest (ca. 1250 m a.s.l.). According to the theory of AAS & FAARLUND (1988) subalpine birch forest, following after the declining pine forest, should have occurred at these localities for enough time to be recorded in the diagrams. However, there is no evidence for this in the *Betula* curve at any of the sites. On the other hand, the results of MOE (1978), SIMONSEN (1980), SELSING (1986), and KVAMME (1989) all show that the subalpine birch forest was established at its present altitudinal range in these areas during the mid-Holocene.

In spite of these arguments, the hypothesis of AAS & FAARLUND (1988) can hardly be said to be rigourously falsified (*sensu* POPPER, 1959). Rather it has been formulated without taking into consideration all the relevant data. The hypothesis is, however, possible to test, but this has not yet been properly done. The subalpine birch forest is one of the most exclusive types of vegetation found in Scandinavia today. Improved understanding of how, when, and why it came into existence is therefore a stimulating task for future research.

Acknowledgements

The author wishes to thank Norges Vassdrags- og Energiverk (NVE) and Statkraft for financial support, Siri Herland for technical help with the figures, Dagfinn Moe and Arvid Odland for providing unpublished materials, Atle Nesje, Knut Fægri, and Peter Emil Kaland for good cooperation and stimulating discussions, and in particular John Birks for scientific inspiration and help with the manuscript.

References

AAS, B. (1964): Bjørke- og barskoggrenser i Norge. Unpublished thesis, Univ. of Oslo

AAS, B. & FAARLUND, T. (1988): Postglasiale skoggrenser i sentrale sørnorske fjelltrakter. Norsk Geogr. Tidsskr. 42, 25-61

BARTH, E. K.; LIMA DE-FARIA, A. & BERGLUND, B. E. (1980): Two ^{14}C dates of wood samples from Rondane, Norway. Bot. Notiser 133, 643-644

BIRKS, H. J. B. (1986): Late Quaternary biotic changes in terrestrial and lacustrine environments, with particular reference to Northwest Europe. In: Berglund, B. E. (ed.): Handbook of Holocene Palaeoecology and Palaeohydrology. J. Wiley & Sons, Chichester, 3-65

BIRKS, H. J. B. (1988): Long-term ecological change in the British uplands. In: Usher, M. B. & Thompson, D. B. A. (eds.): Ecological change in the uplands. Blackwell Scientific Publications, Oxford, 37-56

BIRKS, H. J. B. (1989): Holocene isochrone maps and patterns of tree-spreading in the British isles. J. Biogeogr. 16, 503-540

BIRKS, H. J. B. (1990): Changes in vegetation and climate during the Holocene of Europe. In: Boer, M. M. & de Groot, R. S. (eds.): Landscape-Ecological Impact of Climatic Change. IOS-Press, Amsterdam, 133-157

BORTENSCHLAGER, S. (1982): Chronostratigraphic subdivisions of the Holocene in the Alps. Striae, 16, 75-79

BORTENSCHLAGER, S. (1989): Glacier fluctuations and changes in the forest-limit in the Alps. In: Rupke, J. & Boer, M. M. (eds.): Landscape-ecological impact of climatic change - Discussion Report on the Alpine regions. 88-99, Agricultural Univ. of Wageningen, Univ. of Utrecht, Univ. of Amsterdam

BURGA, C. A. (1988): Swiss vegetation history during the last 18000 years. New Phytol. 110, 581-602

CASELDINE, C. J. & MATTHEWS, J. A. (1987): Podzol development, vegetation change and glacier variations at Haugabreen, South Norway. Boreas 16, 215-230

COHMAP (1988): Climatic changes of the last 18000 years: Observations and model simulations. Science 241, 1043-1052

DANIELSEN, A. (1970): Pollen-analytical Late Quaternary Studies in the Ra district of Østfold, Southeast Norway. Årb. Univ. Bergen 1969, Mat.-Naturv. Serie 14, 1-146.

FÆGRI, K. (1945): A pollen diagram from the subalpine region of Central South Norway. Norsk Geol. Tidsskr. 25, 99-126

HAFSTEN, U. (1960): Pollen-analytical investigations in South Norway. In: Holtedahl, O. (ed.): Geology of Norway. Norsk Geol. Unders. 208, 434-462

HAFSTEN, U. (1965): Vegetational history and land occupation in Valldalen in the subalpine region of Central South Norway traced by pollen analysis and radiocarbon measurements. Årb. Univ. Bergen 1965, Mat.-Naturv. Serie 3, 1-26

HUNTLEY, B. & PRENTICE, I. C. (1988): July temperatures in Europe from pollen data, 6000 years before present. Science 241, 687-690

INDRELID, S. & MOE, D. (1982): Februk på Hardangervidda i yngre steinalder. Viking 36, 36-71

IVERSEN, J. (1960): Problems of the Early Postglacial forest development in Denmark. Danm. Geol. Unders. Ser. IV 4, (3), 1-32

IVERSEN, J. (1973): The development of Denmark's nature since the last glacial. Danm. Geol. Unders. Ser. V, 7-C, 1-126

KULLMAN, L. (1979): Change and stability in the altitude of the birch tree limit in the Southern Swedish Scandes 1915-1975. Acta Phytogeogr. Suec. 65, 121 p.

KULLMAN, L. (1988): Holocene history of the forest-alpine tundra ecotone in the Scandes Mountains (Central Sweden). New Phytol. 108, 101-110

KULLMAN, L. (1989): Tree limit history during the Holocene in the Scandes Mountains, Sweden, inferred from subfossil wood. Rev. Palaeobot. Palynol. 58, 163-171

KULLMAN, L. (1990): Dynamics of altitudinal tree limits in Sweden: a review. Norsk Geogr. Tidsskr. 44, 103-116

KUTZBACH, J. E. & GALLIMORE, R. G. (1988): Sensitivity of a coupled atmosphere/mixed layer ocean model to changes in orbital forcing at 9000 years B.P. J. Geophys. Res. 93, 803-821

KUTZBACH, J. E. & GUETTER, P. J. (1986): The influence of changing orbital parameters and surface boundary conditions on climatic simulations for the past 18000 years. J. Atmosph. Sci. 43, 1726-1759

KVAMME, M. (1984): Vegetasjonshistoriske undersøkelser. In: Meyer, O. B. (ed.): Breheimen-Stryn. Konsesjonsavgjørende botaniske undersøkelser. Botanical Institute, Univ. of Bergen, Report 34, 238-275

KVAMME, M. (1988): Pollen analytical studies of mountain summer farming in Western Norway. In: Birks, H. H., Birks, H. J. B., Kaland, P. E. & Moe, D. (eds.): The cultural landscape - past, present and future. Cambridge Univ. Press, 349-367

KVAMME, M. (1989): Vegetasjonshistoriske undersøkelser i Sprongdalen. In: Odland, A., Aarrestad, P. A. & Kvamme, M. (eds.): Botaniske undersøkelser i forbindelse med vassdragsregulering i Jostedalen, Sogn & Fjordane. Botanical Institute, Univ. of Bergen, Report 47, 166-202

KVAMME, M. (in prep.): Holocene mountain forest history and its relation to local climate and glacier development at Jostedalsbreen, Western Norway

LIESTØL, O. (1960): Glaciers of the present day. In Holtedahl, O. (ed.): Geology of Norway. Norsk Geol. Unders. 208, 434-462

LIESTØL, O. (1969): Brefluktuasjoner. In: Østrem, G. & Ziegler, T. (eds.): Atlas over breer i Sør-Norge. Meddelse Nr. 20 fra Hydrologisk avdeling, Norges vassdrags- og elektrisitetsvesen, 14-16

MANGERUD, J.; ANDERSEN, S. T.; BERGLUND, B. E. & DONNER, J. (1974): Quaternary stratigraphy of Norden, a proposal for terminology and classification. Boreas 3, 109-128

MOE, D: (1973): Studies in the Holocene vegetation development on Hardangervidda, Southern Norway. I. Norwegian Arch. Rev. 6, 67-73

MOE, D. (1978): Studier over vegetasjonsutviklingen gjennom Holocen på Hardangervidda, Sør-Norge. II. Generell utvikling og tregrensevariasjoner. University of Bergen, Univ. of Bergen, Ph. Dr. Thesis, 99 p.

MOE, D. (1979): Tregrensefluktuasjoner på Hardangervidda etter siste istid. In: Nydal, R.; Westin, S.; Hafsten, U. & Gulliksen, S. (eds.): Fortiden i søkelyset. Laboratoriet for Radiologisk Datering, Trondheim, 199-208

MOE, D. & ODLAND, A. (in press): A palaeo- and recent-ecological discussion of *Alnus incana* in Norway. (A contribution to the study of climatic variation through Holocene). Acta Bot. Fenn.

NESJE, N. & KVAMME, M. (in press): Holocene glacier and climatic variations in Western Norway: evidence for Early Holocene glacier demise and multiple Neoglacial events. Geology

NESJE, N.; KVAMME, M.; RYE, N. & LØVLIE, R. (1991): Holocene glacial and climate history of the Jostedalsbreen region, Western Norway: evidence from lake sediments and terrestrial deposits. Quat. Sci. Rev. 10, 87-114

NORDHAGEN, R. (1943): Sikilsdalen og Norges fjellbeiter. Bergens Museums Skrifter 22, 1-607

ODLAND, A.; AARRESTAD, P. A. & KVAMME, M. (1989): Botaniske undersøkelser i forbindelse med vassdragsregulering i Jostedalen, Sogn & Fjordane. Botanical Inst., Univ. of Bergen, Report 47, 166-202

ODLAND, A.; BEVANGER, K.; HANSSEN, O. & REITAN, O. (in press): Fjellskog - biologi og forvaltning. NINA Utredning

PATZELT, G. (1974): Holocene variations of glaciers in the Alps. Colloques Int. Centre Natn. Rech. Scient. 219, 51-59

PATZELT, G. & BORTENSCHLAGER, S. (1973): Die postglazialen Gletscher- und Klimaschwankungen in der Venedigergruppe (Hohe Tauern, Ostalpen). Z. Geomorph. Suppl. 16, 25-72.

PAUS, Aa. (1988): Late Weichselian vegetation, climate and floral migration at Sandvikvatn, North Rogaland, Southwestern Norway. Boreas 17, 113-140

POPPER, K. R. (1959): The logic of scientific discovery. Hutchinson, London

PORTER, S. C. & OROMBELLI, G. (1985): Glacial contradiction during the middle Holocene in the western Italian Alps: Evidence and implications. Geology 13, 296-298

ROALD, I. (1973): Litt om sambandet mellom værparametre målt ved breene og på faste værstasjoner i nærheten. In: Tvede, A. (ed.): Glasiologiske undersøkelser i Norge 1971. Hydrologisk avdeling, Vassdragsdirektoratet, Norges Vassdrags- og elektrisitetsvesen, Rapport 2/73, 90-99

SELSING, L. (1986): The first human impact and its relationship to the time of deglaciation and the forest-limit variations in the mountain areas in Southern Norway. Striae 24, 137-142

SELSING, L. & WISHMAN, F. (1984): Mean summer temperatures and circulation in a Southwest Norwegian mountain area during the Atlantic period, based upon changes of the alpine pine-forest limit. Ann. Glaciol. 5, 127-132

SIMONSEN, A. (1980): Vertikale variasjoner i Holocen pollen sedimentasjon i Ulvik, Hardanger. AmS-Varia 8, 1-68

VE, S. (1930): Skogstrærnes forekomst og høydegrenser i Årdal. Medd. Vestl. Forstl. Forsøksstn. 13, 1-94.

Address of the author:

Dr. M. Kvamme, Botanical Institute, University of Bergen, Allégaten 41, N-5007 Bergen, Norway

Holocene timberlines and climate in North Norway - an interdisciplinary approach

K. D. Vorren, C. Jensen, R. Mook, B. Mørkved & T. Thun

Summary

A newly initiated study of the Holocene timberline climate development in Northern Norway is reported here. To achieve an optimal interpretation of Holocene climatic proxy data, overlapping chrono-segments of physical-meteorological observation series, tree-ring series, and stratigraphical sequences at the timberline, will be studied and compared. A test study seems to indicate that influx pollen data from sediments at different altitudinal levels below and above present timberline might be used for detecting the climatically conditioned timberline oscillations during the Holocene.

Zusammenfassung

In Nord-Norwegen ist eine Untersuchung über die Entwicklung des holozänen Waldgrenzenklimas eingeleitet worden. Um die klimatischen Proxydaten optimal auswerten zu können, werden sich überlappende Zeitreihen von physikalisch-meteorologischen Messungen, Kiefernholz-Jahrringen und stratigraphischen Sequenzen (Torf- und Seesedimente) aus der Waldgrenzregion aufbereitet und verglichen. Eine Pilotstudie scheint anzudeuten, daß Polleninfluxdaten auf Sedimentpaketen ober- und unterhalb der Waldgrenze holozäne, klimatisch bedingte Waldgrenzfluktuationen anzeigen können.

1. Introduction

The timberline, or more appropriately, the upper limit of woodland in Scandinavia is formed by birch (*Betula pubescens* Ehrh.s.l./*Betula tortuosa*). However, in many areas there is also a clear upper pine timberline, which in Northern Norway is formed by *Pinus sylvestris* L. ssp. *lapponica*, some 200-300 m below the orographically conditioned upper *Betula* limit. Both these limits may have oscillated during the Holocene, but not necessarily synchronously.

To relate the forest limits and their oscillations to climate, a small-scale investigation in the middle part of Northern Norway (68°47'-69°22' N and 15°45'-20°25' E) has been initiated.

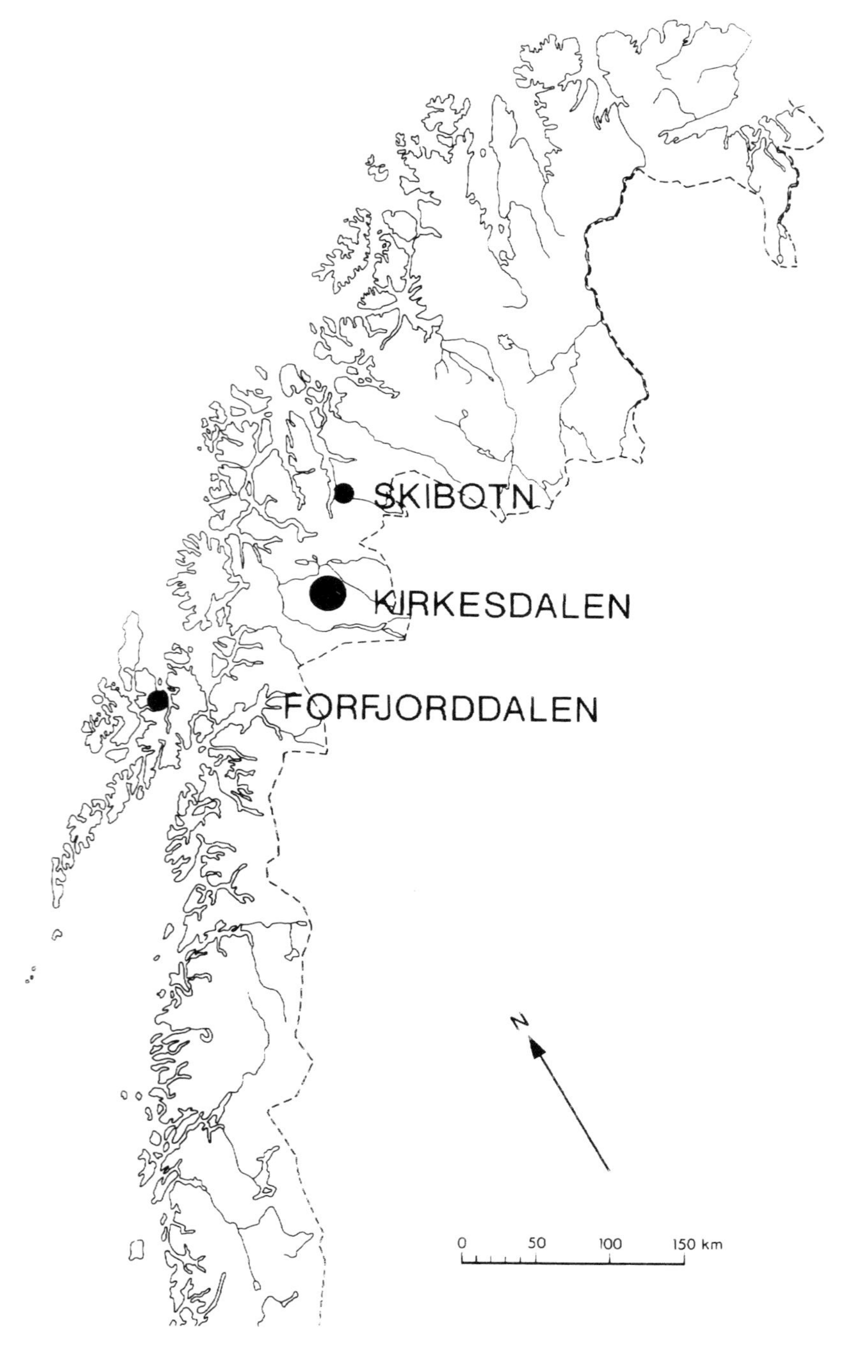

Fig. 1 Investigation sites, in the districts of Tromsø and Nordland, North Norway

1.1 Investigation sites

Data are being collected from three different site areas (Fig. 1):

(1) Skibotn, ca. 69°22' N, 20°25' E. Climate is sub-continental with a mean annual lowland temperature amplitude of ca. 23K (average temp. July ca. 14°C and for February ca.-9°C. Annual precipitation averages 350-400 mm). From Skibotn, recent timberline temperature data and tree-ring data will be collected. Comparative sediment data can be collected from small mires;

(2) Kirkesdalen, ca. 68°45'-69°N, 18°30'-19°30' E. Climate is sub-maritime to subcontinental with an annual average lowland temperature amplitude of 20-23K, and precipitation probably between 300 and 500 mm. From Kirkesdalen, tree-ring and sediment data will be collected. In this region both lakes and mires occur relatively frequently at suitable altitudes;

(3) Forfjorddalen, 68°47' N, 15°45' E, a site within a suboceanic vegetation zone (with *Blechnum-Narthecium* vegetation). The annual average temperature amplitude here is estimated at ca. 13-14K.). Tree-ring series will be constructed from Forfjorddalen. The age of living *Pinus sylvestris* here amounts to more than 600 years. The *Pinus* enclave of Forfjorddalen is at the cool-maritime border of the *Pinus sylvestris* area.

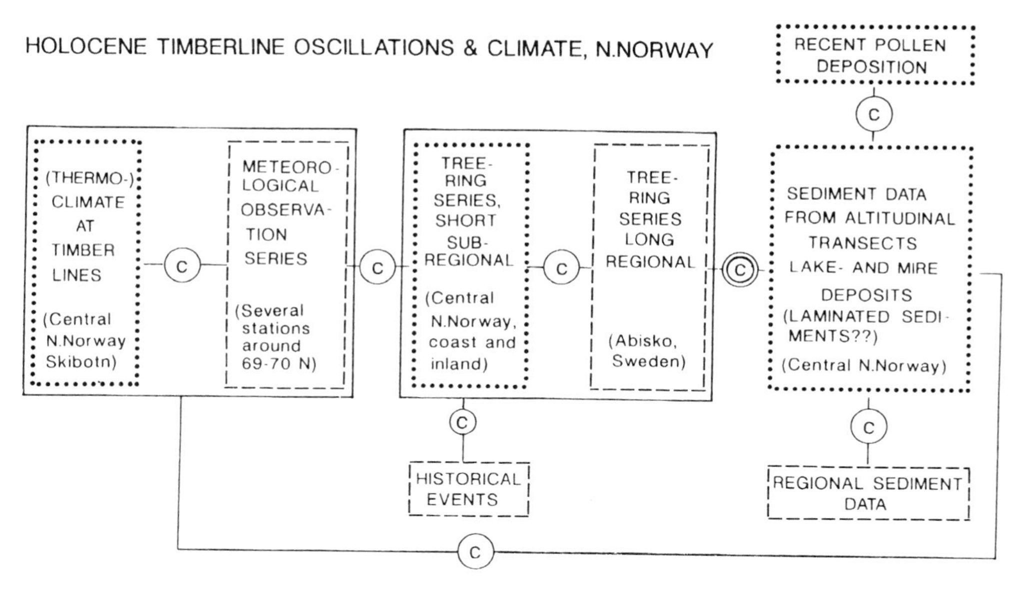

Fig. 2 Project draft, schematic

1.2 Project outline

The main idea of the present project is to create overlapping series of (a) real climate data and (b) climatic proxy data (Fig. 2). The main aim is to use knowledge of the present orographic timberline climate as a basis for the interpretation of proxy data from tree-rings and sediments.

To connect the different data series, one needs a series of calibrations, or transfer functions, established by means of numerical procedures.

The prevailing assumption that temperature parameters express the total heat balance of the orographic vegetation limits, and especially the timberlines, was the premise for a temperature climate project in Skibotn (MOOK & VORREN, 1990). Four continuously logging stations (sensors at different levels above and below ground) have been established at the most important orographic vegetation belt limits, of which the two lowermost are the *Betula* and the *Pinus* limit (Fig. 3 and 4). Temperature data from these stations will be correlated with "standard normals" from the official meteorological observation station network.

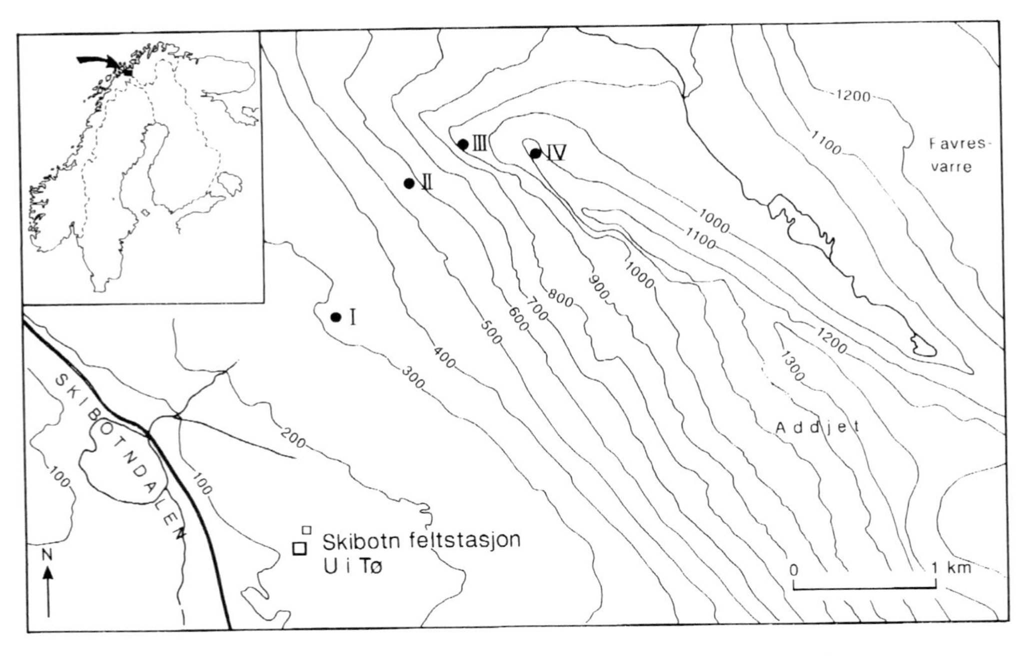

Fig. 3 Position of automatic temperature climate stations in Skibotn (Aanderaa system): I, at the upper orographic *Pinus*-limit; II, at the upper *Betula*-limit; III, at the low-/mid-"alpine" limit (= upper limit of *Betula nana-Empetrum*-heaths, and of *Vaccinium myrtillus*); IV, in the upper part of the mid-"alpine" belt (*Cassiope tetragona*-heaths)

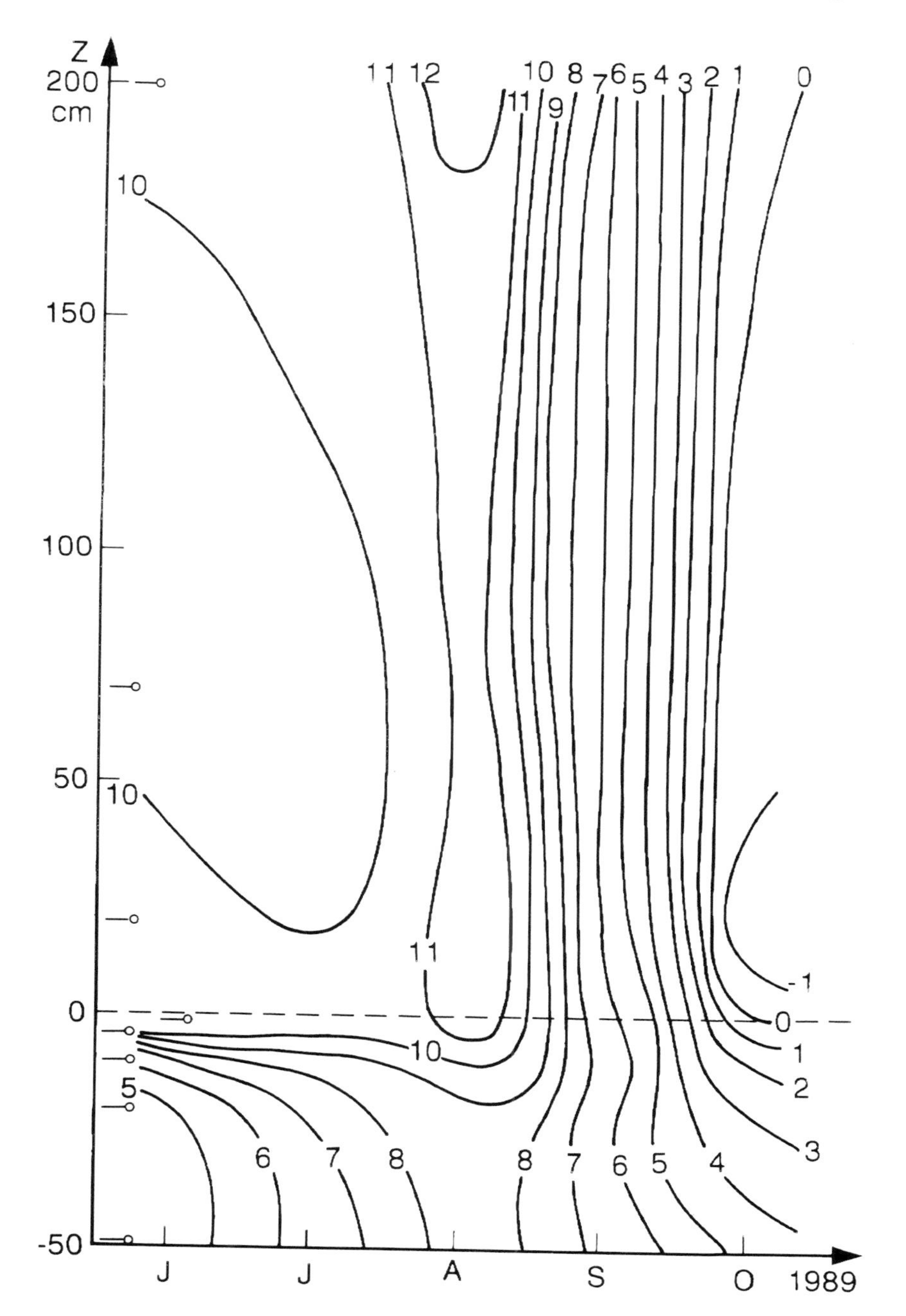

Fig. 4 Temperature climate at station I, Skibotn, upper *Pinus*-timberline, ca. 310 m a.s.l., June-Oct. 1989. Level of 8 temperature sensors (3 above, 1 in, and 4 below soil surface) have been marked

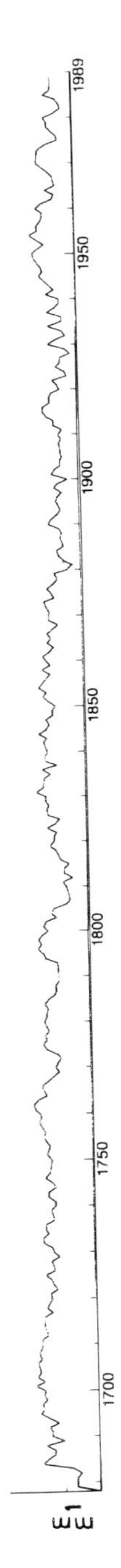

Fig. 5 Tree-ring chronology of *Pinus sylvestris* at the *Pinus*-timberline in the northern part of Kirkesdalen. Based on 7 more or less equally old pines

The next step towards palaeoclimate is to link temperature parameters with sub-regional tree-ring series (Fig. 5 and 6). The annual tree-rings and/or the "summer wood"/maximum late-wood density have to be correlated with adequate temperature parameters like aggregates of monthly means. Based on such correlations we hope to prolong the "standard normal" temperature curve 300 years or more back in time, and maybe also to detect differences between the maritime and the continental climate. The aim is to create three such *Pinus* tree-ring series extending beyond the Little Ice Age, but it may prove difficult to get enough material for more than ca. 300 years back in time.

Being aware of what splendid palaeoclimatic tools the long Abisko tree-ring series (BARTHOLIN & KARLÉN, 1983; SCHWEINGRUBER et al., 1988; BRIFFA et al. 1990), and the joint one prepared for Finnish Lapland and Swedish Lapland (ERONEN & HUTTUNEN, 1990) may turn out to be, we consider it vital to compare our short series with those mentioned by means of current mathematical methods.

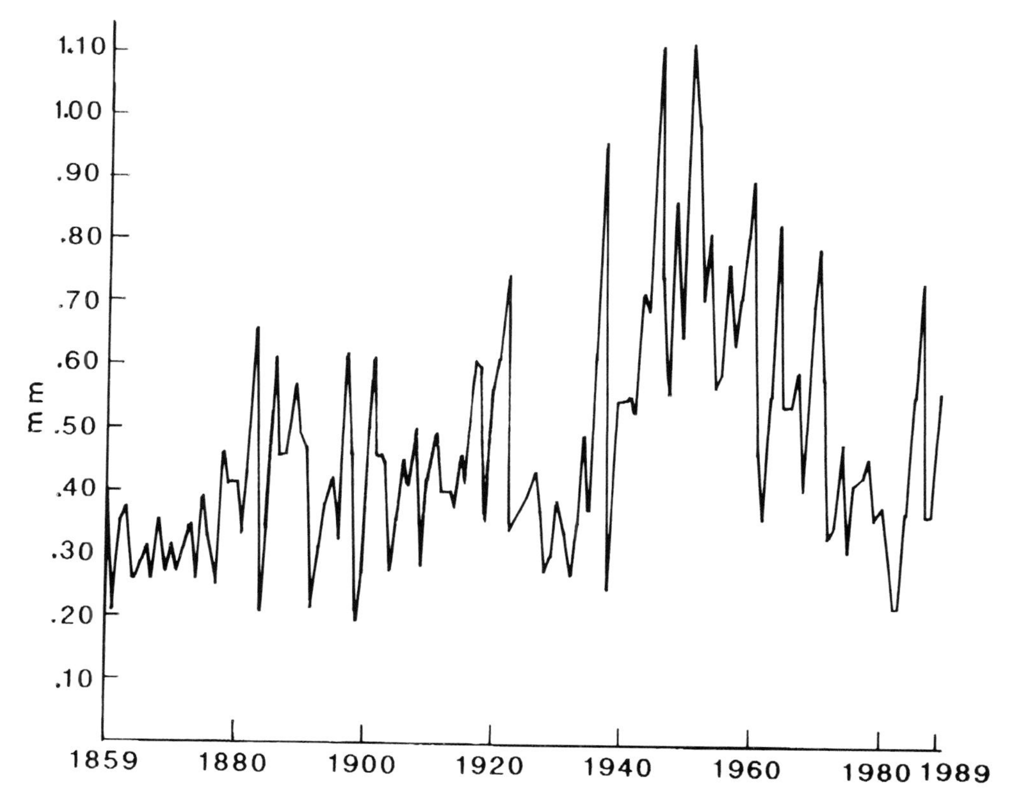

Fig. 6 Summer wood (wood with maximum density); measures assessed by optic instruments. Based on 9 more or less equally old pines from a southwest-exposed slope near the valley floor, northern part of Kirkesdalen

The missing link between tree-ring data and current sediment data in our region so far, is annually laminated sediments. A search for such sediments is going on, but they seem to be rare in our humid region, at least at the timberline level.

For comparison between subfossil and recent pollen deposition in vertical transects along mountain slopes, we wish to use Tauber traps in a combination with moss samples, especially tight-growing Sphagna Acutifolia and *Sphagnum angustifolium*.

The need for a detailed ^{14}C-chronology cannot be emphasized strongly enough. To secure safe datings, wet-sieving of peat, which selects moss lamina and excludes coarser fragments that may be of a more recent origin, should be carried out. Unless special fundings for commercial datings are established, the procedure of radiocarbon datings will unavoidably mean a delay of the total project.

A detailed, continuous analysis of other lake sediments, peats and alluvial sediments at the timberline region may not be expected to give a higher resolution than 40-60 years with current methods and sediment thicknesses. Hiatuses may be expected in mountainous sediments, due to solifluction and other cryospheric erosion processes.

In the different sediment types, both lithological and biological parameters will be analysed with short vertical intervals or continuously in centimetre steps. Normally, a sub-sample integrates sediments of 1 vertical cm thickness, but it is possible to reduce this thickness, if required, and if the sediments are homogeneously and finely decomposed.

2. Stratigraphical test cases

Because of topographical advantages, the Kirkesdal area has been selected as the main area for sediment studies. In a lateral valley, two sites ca. 200 m above the present *Betula* timberline are being investigated, one mire and one lake deposit.The peat deposit (Fig. 7) has loss on ignition values mainly between 20 and 30 percent of dry weight. The curve oscillates between more or less mineral-rich layers. The less mineral-rich layers seem to correlate with *Pinus*-rich periods. Probably the mineral-rich layers are connected with periods of greater humidity and more meltwater from the environments, and the mineral-poor layers with warm and dry periods. However, local topography, especially stream courses, may change throughout the times, a circumstance that may complicate the interpretation of the loss on ignition curve. The AP-concentration and influx diagrams show a lower main period between ca. 8000 and 4000 yr B.P. with high concentration and influx values, and an upper part with lower values. Compared with data from Eastern Finnmark (HYVÄRINEN, 1975), this is at least an indication of an Early or mid-Holocene higher timberline and a Late Holocene lower timberline. The chronological dimension of the influx curve fits very well with the one obtained for subfossil tree remains above the present timberline, presented by ERONEN & HUTTUNEN (1990). Additional recent pollen deposition studies, sediment studies from sites above and below the present one, and macrofossil studies may give decisive evidence of the presence or absence of woodland at the 800 m level during the Early and mid-Holocene. It is interesting that only the uppermost layers, after 1600 B.P., have influx figures for AP below 200, whereas there are periods also after ca. 4000 yr B.P. with influx figures close to those between 8000 and 4000 B.P. (800-900 pg x cm^{-2} x yr^{-1}). HICKS (1990) has similar or slightly higher figures for AP in *Betula*-forest near its altitudinal limit in Finnish Lapland, and HYVÄRINEN (1975) has figures below these from subfossil sequences near the arctic *Betula*-woodland border in Northeast Norway.

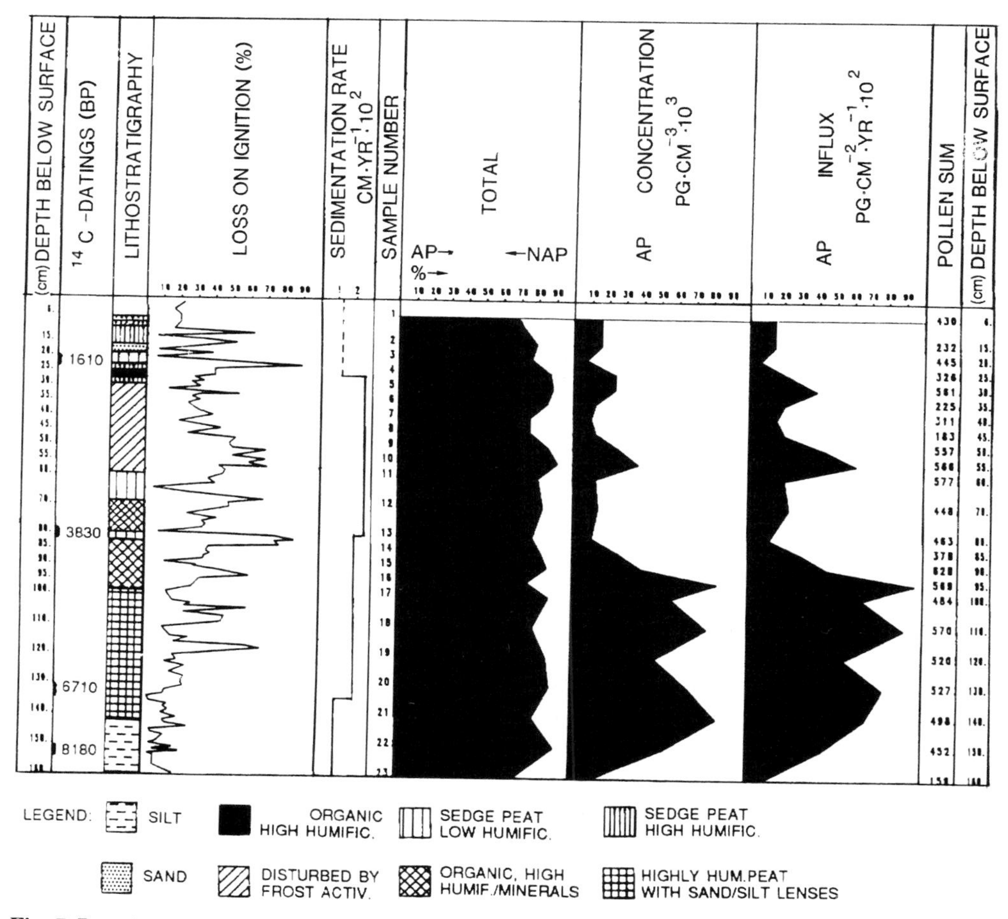

Fig. 7 Peat deposit at ca. 800 m a.s.l. in the eastern part of Lappskardet, upper Kirkesdalen, ca. 200 m above the present upper *Betula*-limit: loss on ignition, total pollen, AP-concentration, AP-influx. The ^{14}C-datings are, from top to bottom: Beta-39466:1610±80, Beta-39467: 3830±90, Beta-39468: 6710±140, Beta-39469: 8180±240

The lake sediments are much poorer in organic components (5-10 %), due to a variety of nival and cryogenic processes around the lake (all ^{14}C-datings, Beta-39470 to 39473, are erroneous; according to the dating laboratory Beta Analytic Inc., Florida, most likely due to influx of old, Weichselian (?), organic components). However, here also, a shifting between periods with more stable lake environments, and periods with more instable environments may have occurred. The first periods may be recognized through more or less laminated sediments, and a slightly higher amount of organic matter (Fig. 8). The later periods may be recognized by high concentrations of mineral matter, even layers with drop stone, and little or no lamination.

When synthesizing the palaeoclimatical proxy data, one should consider all relevant stratigraphical data from the region, regardless of whether they are from the timberline or the valley floor. The major vegetation changes as well as the lithological changes recorded in sediments in these northern areas often seem to be related to pronounced climatic events.

For instance, preliminary investigations in North Norwegian ombrotrophic peats sequences have shown a humification chronology rather similar to the one of Danish bogs (NILSSEN & VORREN, 1991).

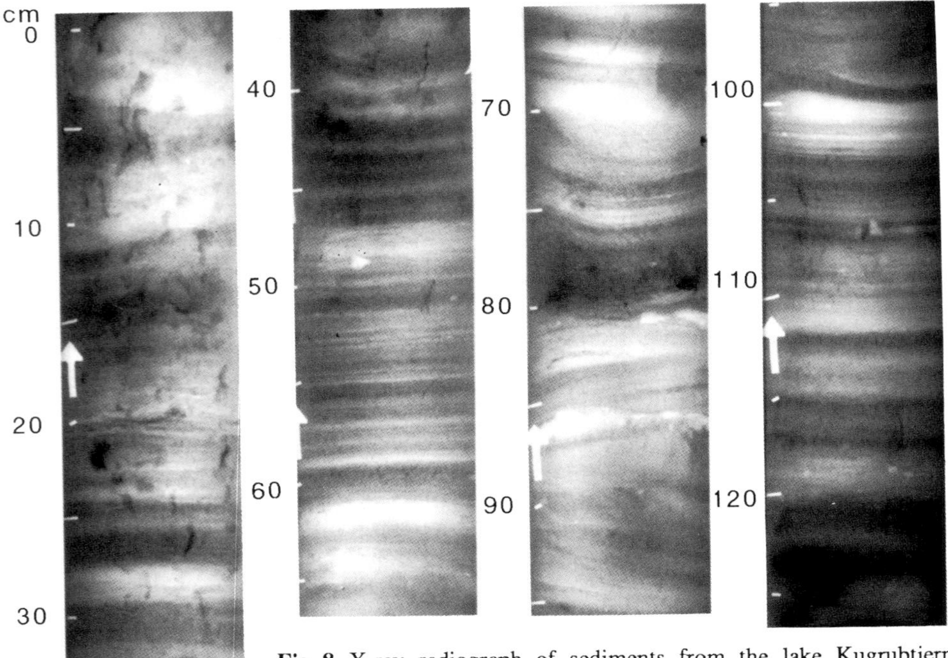

Fig. 8 X-ray radiograph of sediments from the lake Kugrubtjern (or Rundvatnet), ca. 800 m a.s.l. in the northeastern part of Lappskardet, southern part of the Kirkesdalen valley, ca. 200 m above the present upper *Betula*-limit. The 1.3 m deep sediments are partly laminated, and very poor in organogenic matter (5-10% loss on ignition)

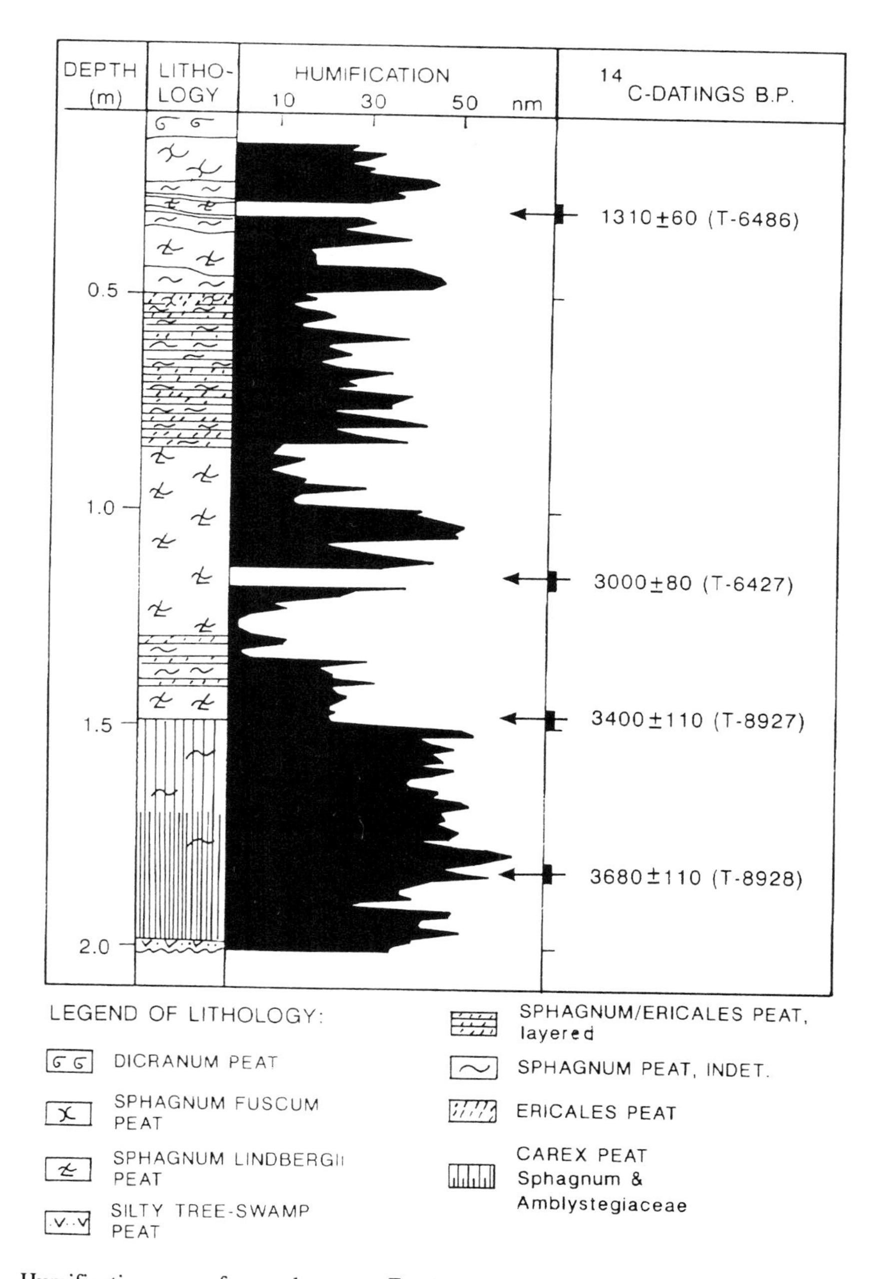

Fig. 9 Humification curve from a bog near Fosshaug/Smerud, Målselv, ca. 6 km northwest of the northern part of Kirkesdalen valley. The sequence probably has a hiatus just above the level dated at 1310 yr B.P. Humification oscillations indicate changes between humid and less humid periods

Ombrotrophic bog peats are not generated at the *Betula*-timberline, but occur at or below the *Pinus sylvestris*-timberline (Fig. 9). When establishing a transect of sites across these two orographic timberlines, it is thus also possible to compare atmospheric humidity as expressed in ombrogenic peat humification, and nival processes related to humidity, in more or less minerogenic sediments, expressed as loss on ignition values.

Acknowledgements

The authors want to express their gratitude to the Norwegian Research Council for the Humanities and Sciences, and to the Comittee of Climate and Ozon Research for financing the present pilot project within the Norwegian EPC. We are much obliged to Geoffrey Corner, University of Tromsø, who corrected the English language.

References

BARTHOLIN, T. & KARLÉN, W. (1983): Dendrokronologi i Lappland. Dendrokronologiska Sällskapet, Meddelanden 5, 1-16

BRIFFA, K.; BARTHOLIN, T. S.; ECKSTEIN, D.; JONES, P. D.; KARLÉN, W.; SCHWEINGRUBER, F. H. & ZETTERBERG, P. (1990): A 1400-year tree-ring record of summer temperatures in Fennoscandia. Nature 346, 434-439

ERONEN, M. & HUTTUNEN, P. (1993): Pine megafossils as indicators of tree limit changes in Fennoscandia. (this volume)

HICKS, S. (1993): The use of recent pollen rain records in investigating natural and anthropogenic changes in the polar tree limit in Northern Fennoscandia. (this volume)

HYVÄRINEN, H. (1975): Absolute and relative pollen diagrams from northernmost Fennoscandia. Fennia 142, 23 p.

MOOK, R. & VORREN, K.-D. (1990): Temperaturklimaet ved grensen mellom fjellvegetasjonsbeltene. (Temperature climate at the transition between orographic vegetation belts). Polarflokken 14/1, 55-108

NILSSEN, E. & VORREN, K.-D. (1991): Peat humification and climate history. Norsk Geol. Tidsskr. 90/32, 215-217

SCHWEINGRUBER, F. H., BARTHOLIN, T., SCHÄR, E. & BRIFFA, K. R. (1988): Radiodensitometric-dendroclimatological conifer chronologies from Lapland and the Alps (Switzerland). Boreas 17, 559-566

Addresses of the authors:

Prof. Dr. K.-D. Vorren, University of Tromsø, N-9000 Tromsø, Norway
Dr. C. Jensen, University of Tromsø, N-9000 Tromsø, Norway
Dr. R. Mook, University of Tromsø, N-9000 Tromsø, Norway
Dr. B. Mørkved, University of Tromsø, N-9000 Tromsø, Norway
Dr. T. Thun, University of Trondheim, N-7050 Dragvoll, Norway

Climate and physiology of trees in the Alpine timberline regions

Walter Tranquillini

Summary

The temperature decrease with altitude in connection with the shortening of the vegetation period lowers the yearly sum of photosynthesis of trees at the timberline. It reduces the increment of all organs, i.e. length and diameter of shoots and roots, and of reproduction organs. It also prohibits the maturation of newly developing organs (shoots, leaves, and seed). The limit of existence for trees in the timberline ecotone is determined by frost damage in beech and desiccation damage in tree species of the Central Alpine timberline.

Zusammenfassung

Die Temperaturabnahme mit der Höhe und die damit verbundene Verkürzung der Vegetationsperiode erniedrigt die Jahressumme der Photosynthese der Bäume an der Waldgrenze, verringert den Zuwachs aller Organe (Länge und Dicke der Sprosse und Wurzeln, Fruktifikationsorgane) und verhindert die Ausreifung der neugebildeten Organe (Triebe, Blätter, Samen). Die Existenzgrenze der Bäume in der Kampfzone wird bei der Buche durch Frostschäden, bei den Baumarten der zentralalpinen Waldgrenze durch Austrocknungsschäden bestimmt.

1. Introduction

There is ample evidence of a causal connection between the altitude of the upper forest limit and the climate. If the climate changes the timberline will also drift upwards or downwards.

There are several indications that temperature plays an essential role in the upper limit of the distribution of the forest, as can be shown for instance in the decrease in altitude of the timberline from the Subtropic to the Arctic region (DAUBENMIRE, 1954) or, on a smaller scale, in the higher timberline on warm southern slopes compared to cool northern slopes (KÖSTLER & MAYER, 1970).

From this, one may conclude that it is mainly changes in temperature that lead to movements of the timberline. On the other hand one may deduce changes of the temperature climate in the past from known shifts of the timberline.

However, the effect of temperature on plants is very complex including direct influence on numerous physiological processes. Temperature alters the physiological conditions of plants and thereby their response to climatic factors. Very high and low temperatures ultimately lead to heat and frost damage and in the extreme to the death of the plants (TRANQUILLINI, 1979a).

In this paper I would like to analyse some important effects of temperature on trees on an ecophysiological basis.

2. Photosynthesis

2.1 The temperature dependence of net photosynthesis

The relationship between photosynthesis and temperature has the form of a curve rising to an optimum and then declining. According to this optimum curve we thought at first that temperature, which decreases with altitude, must lead to a decrease in the photosynthetic rate (TRANQUILLINI & TURNER, 1961). Today we know that this effect is kept small by genetic and modificative adaptations of photosynthesis (SLATYER, 1977). Provenances from the timberline have a lower temperature optimum of photosynthesis than provenances from low elevations and can even reduce it further if they are exposed to low temperatures for some days (Fig. 1).

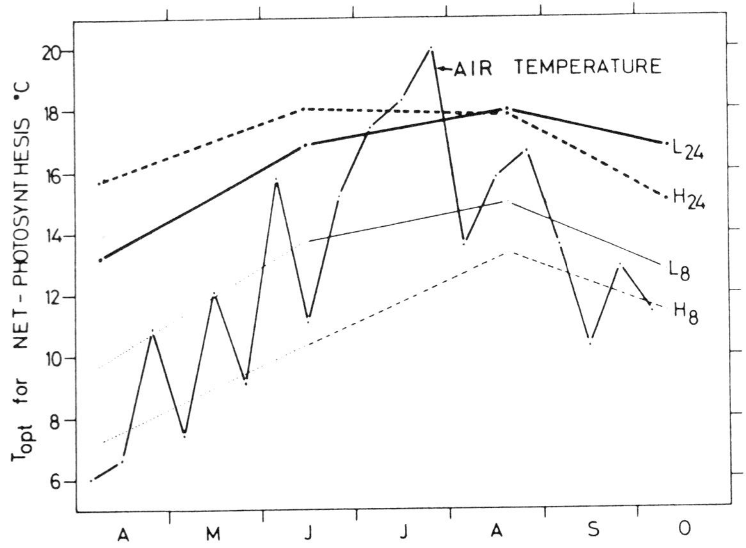

Fig. 1 Optimum temperature for photosynthesis after pre-treatment at 8°C or 24°C at different times of the year of spruce seedlings of low (L) and high (H) altitude provenances grown in a nursery at 1100 m a.s.l. For comparison mean air temperatures (10 day means) from a meteorological station near the nursery are shown. (From TRANQUILLINI & HAVRANEK, 1985)

2.2 The duration of the period of photosynthetic activity

In evergreen trees the time the needles are photosynthetically active becomes shorter with increasing altitude (PISEK & WINKLER, 1958). For spruce at the Central Alpine timberline the period for photosynthesis was about 166 days in the mean of 10 years (HAVRANEK, 1986). In deciduous trees the length of the foliation period diminishes with increasing altitude, because at high elevations they flush later and turn yellow and shed leaves earlier than at low elevations. At low elevations the foliation period of the early flushing larch is considerably longer than that of beech. However, at their respective altitudinal limit of occurrence it is about the same in both species (Table 1).

Table 1 Length of the foliation period = length of the photosynthetic period (bud break - senescence) of larch (*Larix decidua*) needles and beech (*Fagus sylvatica*) leaves in days in relation to altitude

Altitude	**Length of foliation period**				
m (a.s.l.)	*Larix*	*Fagus*			
	North Tyrol Central Alps	South Tyrol		North Tyrol Limestone Alps	
	mean values of 6 years	1981	1982	1981	1982
700	223	165	166	155	158
1000	214				
1500	183	128	140	110	134
1630		117	128	treeline	
2000	145	treeline			
2100	128				
	treeline				

3. Growth and development

3.1. Height growth in relation to temperature

With increasing altitude the height and diameter growth of all tree species becomes smaller (Fig. 2). This decrease is caused both by an increasingly lower growth rate and by the shortening of the period of extension growth. In tree species belonging to the *Quercus* type with fully preformed shoots in the winter bud (HOFFMANN & LYR, 1973) like Swiss stone pine, Norway spruce, and beech, growth rate is more important for the reduction of the height growth than the duration of the extension growth, whereas in tree species belonging to the *Populus* type with free growth, such as European larch, the duration of extension growth is most important (TRANQUILLINI & PLANK, 1989).

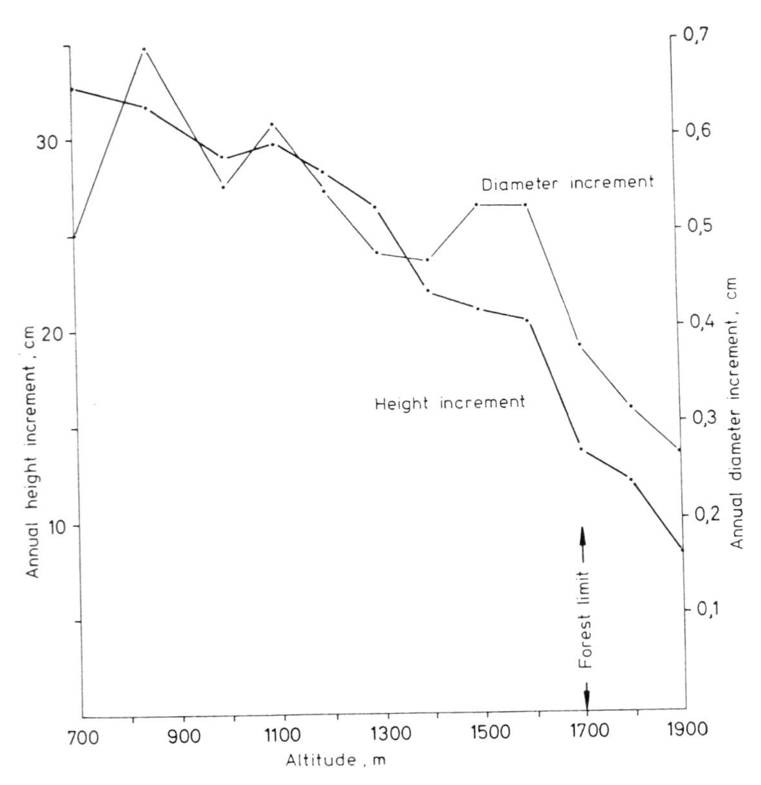

Fig. 2 Mean annual increments of height and stem diameter of *Picea abies* in the Seetal Alps, Austria. 20 mature trees were selected at altitudinal intervals of 100 m and mean growth rates were calculated from tree age and total tree height and diameter respectively. (From HOLZER, 1973)

3.2 Maturation of the new flush and winter desiccation

At the end of the extension growth and during the setting of the terminal bud the phase of maturation takes place. The cuticle and the cutinization of epidermal cells in needles and axes develop and epicuticular wax extrudes. With advancing maturation the cuticular transpiration resistance increases and hence the cuticular transpiration decreases (Fig 3 A). We know that spruce needles in low elevations need three months for the full development of the protective tissues against evaporative water loss (LANGE & SCHULZE, 1966). At the timberline this development may be slower according to the prevailing low temperatures there and thus require more time for the process of maturation. Due to the shortness of the warm season at the timberline the maturation cannot be completed, especially in years with cool summers. Consequently the young shoots meet winter in an immature state. Under conditions of frost drought, which means there is no water supply from the soil or from the frozen water reservoirs in stem and twigs yet high evaporative demand with strong radiation and high needle temperatures, these immature shoots will develop greater water deficits than better ripened shoots at low elevations (BAIG et al., 1974). In unfavourable years they suffer desiccation damages in the late winter (Fig. 3 C), especially at the krummholz limit in the continental part of the Alps and in the Rocky Mountains (TRANQUILLINI 1982; HADLEY & SMITH, 1986).

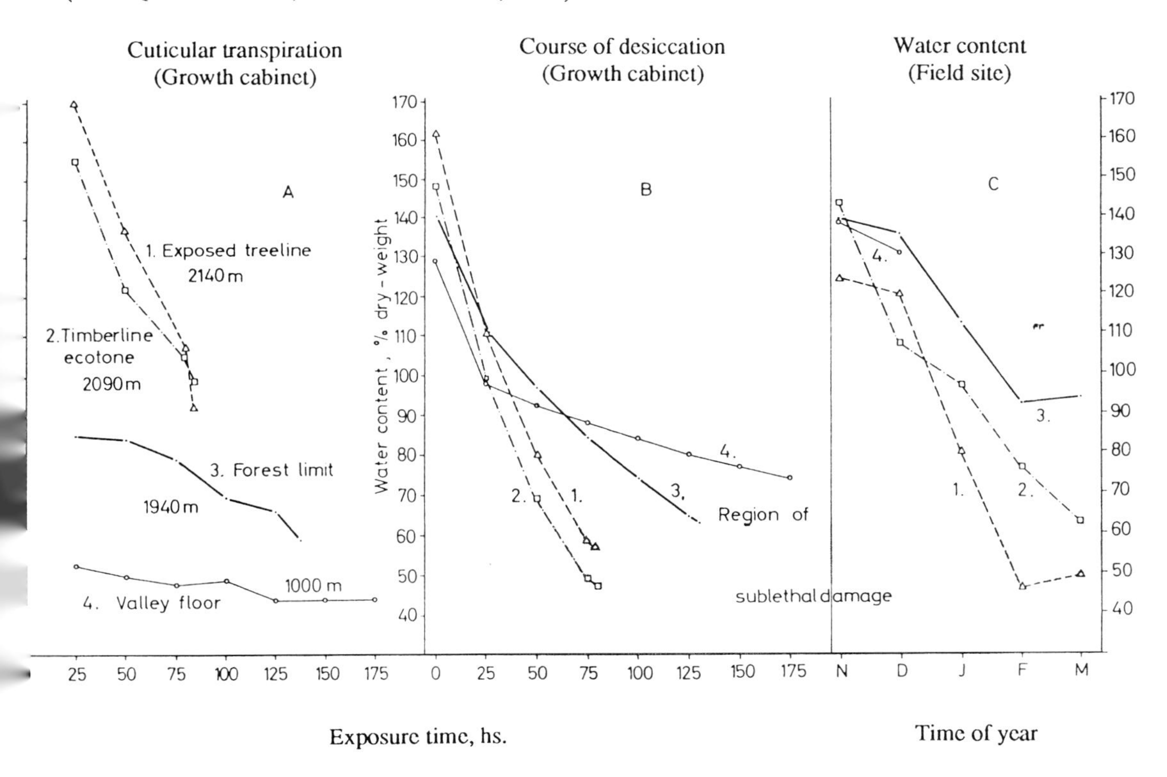

Fig. 3 A) Water loss through cuticular transpiration in excised current year shoots of *Picea abies* in a climatised chamber during winter. Shoots originated from the valley (1000 m), forest limit (1940 m), timberline ecotone (2090 m) and the krummholz limit (2140 m) on Patscherkofel near Innsbruck. B) Change in water content of shoots during progressive desiccation in a climatised chamber. Critical water content is hatched. C) Water content of *Picea abies* shoots from trees at different altitudes on Patscherkofel during winter 1972/73 (from BAIG et al., 1974)

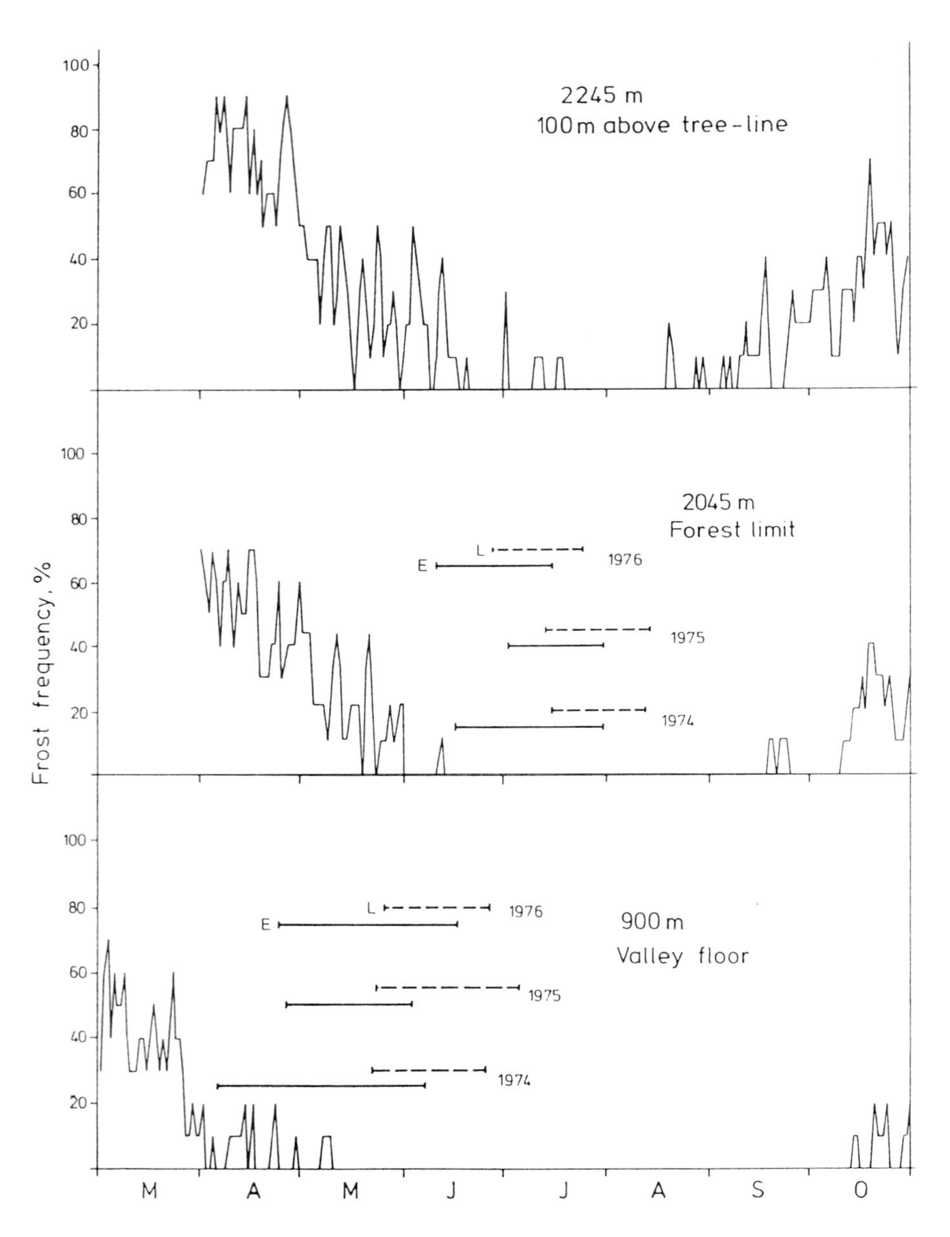

Fig. 4 Daily frequency of temperatures below -3°C during the growing season for the years 1957-1966 at the weather stations Rinn, 900 m (valley floor), Patscherkofel, 2045 m (near forest limit), and Patscherkofel summit, 2245 m (100 m above krummholz limit), Austria. For the years 1974-1976 the growth period of terminal shoots is shown for early (E) and late flushing (L) clones of spruce (*Picea abies*), which were grown at 900 and 2045 m a.s.l. respectively. (From TRANQUILLINI, 1979b).

4. Frost resistance

4.1 Frost resistance in winter

In mid-winter the minima of air temperature near the timberline are only slightly lower than in the valley because of the frequent temperature inversions and the formation of a warm zone along the slopes of the valley which can extend right up to the timberline. However, the decrease in temperature with altitude results in an earlier occurrence of frost in autumn and longer lasting frost events in spring compared to low elevations. In general the seasonal development of the frost resistance is so well adapted to the seasonal course of the temperature that even in transition periods tree species of the Central Alpine timberline keep a wide margin between temperature minimum and frost resistance which is always more than 12 K according to GROSS et al. (1991).

Another pattern is found in beech at the timberline in the Northern Limestone Alps. The leafless shoots of beech are much more frost sensitive than those of conifers. The maximum frost resistance of their buds is only -27°C, that of the axes -33°C, while the resistance of spruce needles is -44°C. At the timberline buds of beech are therefore in danger of frost damage both in mid-winter and during the dehardening period in spring (TRANQUILLINI & PLANK, 1989).

4.2 Frost resistance during the flushing period

There is also a developmental stage in evergreen conifers during which these winter frost resistant trees suffer frost damage from as little as -3°C: it is during the extension growth of the new shoot (SCHWARZ, 1970). This frost sensitive state lasts about 50 days in spruce. At the timberline it fits exactly into that time of the season during which temperatures stay safely above this value (Fig. 4). With increasing altitude the period without damaging frosts becomes shorter and shorter and is interrupted by single frost events. This would make a normal development of the flush impossible. However, in general spruce does not extend into this summer-frost endangered altitudinal region, because spruce trees already reach their limit of existence at lower elevations due to frost drought (TRANQUILLINI, 1979b).

5. Natural regeneration

The altitudinal level of the timberline is also determined by the capability to develop and to distribute germinable seed. With altitude both the frequency of cone years decreases and seed numbers and seed vitality decline. Often the shortness of the vegetation period restricts the development of cones and seed and prohibits seed maturation (SHEARER, 1986). The highest region in which trees produce germinable seed is still below the timberline.

Seed is transported either by birds (for instance in Swiss stone pine) or by the wind. This transport is limited to distances of 50 m for spruce (KUOCH, 1965) and 100 to 200 m for Swiss stone pine (MATTES, 1982).

References

BAIG, M. N.; TRANQUILLINI, W. & HAVRANEK, W. M. (1974): Cuticuläre Transpiration von *Picea abies*- und *Pinus cembra*-Zweigen aus verschiedener Seehöhe und ihre Bedeutung für die winterliche Austrocknung der Bäume an der alpinen Waldgrenze. Cbl. f. d. ges. Forstwesen 91, 195-211

DAUBENMIRE, R. (1954): Alpine timberlines in the Americas and their interpretation. Butler Univ. Bot. Stud. 11, 119-136

GROSS, M.; RAINER, I. & TRANQUILLINI, W. (1991): Über die Frostresistenz der Fichte mit besonderer Berücksichtigung der Zahl der Gefrierzyklen und der Geschwindigkeit der Temperaturänderung beim Frieren und Auftauen. Forstwiss. Cbl. 110, 207-217

HADLEY, J. L. & SMITH, W. K. (1986): Wind effects on needles of timberline conifers: Seasonal influence on mortality. Ecology 67, 12-19

HAVRANEK, W. M. (1986): Physiologische Reaktionen auf Klimastreß bei Bäumen an der Waldgrenze. Proceed. Sympos. Klima und Witterung in Zusammenhang mit den neuartigen Waldschäden. Ges. Strahlen Umweltforschg. München, Ber. 10/87, 115-130

HOFFMANN, G. & LYR, H. (1973): Charakterisierung des Wachstumsverhaltens von Pflanzen durch Wachstumsschemata. Flora 162, 81-98

HOLZER, K. (1973): Die Vererbung von physiologischen und morphologischen Eigenschaften der Fichte. II. Mutterbaummerkmale. Unveröffentlichtes Manuskript

KÖSTLER, J. N. & MAYER, H. (1970): Waldgrenzen im Berchtesgadener Land. Jahrb. Ver. Schutze Alpenpfl. Tiere, München 35, 1-35

KUOCH, R. (1965): Der Samenanfall 1962/63 an der oberen Fichtenwaldgrenze im Sertigtal. Mitt. Schweiz. Anst. Forstl. Versuchsw. 41, 63-85

LANGE, O. L. & SCHULZE, E. D. (1966): Untersuchungen über die Dickenentwicklung der kutikularen Zellwandschichten bei der Fichtennadel. Forstwiss. Cbl. 85, 27-38

MATTES, H. (1982): Die Lebensgemeinschaft von Tannenhäher und Arve. Eidg. Anst. Forstl. Versuchswes. Ber. 241, 74 p.

PISEK, A. & WINKLER, E. (1958): Assimilationsvermögen und Respiration der Fichte (*Picea excelsa* Link) in verschiedener Höhenlage und der Zirbe (*Pinus cembra* L.) an der alpinen Waldgrenze. Planta 51, 518-543

SCHWARZ, W. (1970): Der Einfluß der Photoperiode auf das Austreiben, die Frosthärte und die Hitzeresistenz von Zirben und Alpenrosen. Flora 159, 258-285

SHEARER, R. C. (1986): Effects of elevation on *Pseudotsuga menziesii var. glauca* cone and seed maturity in Western Montana, USA. Forstl. Bundesvers.anst. Wien, Ber. 12, 79-90

SLATYER, R. O. (1977): Altitudinal variation in the photosynthetic characteristics of snow gum, *Eucalyptus pauciflora Sieb. ex Spreng*. III. Temperature response of material grown in contrasting thermal environments. Aust. J. Plant Physiol. 4, 301-312

TRANQUILLINI, W. (1979a): Physiological ecology of the Alpine timberline. Ecol. Stud. 31, Springer, Berlin, 137 p.

TRANQUILLINI, W. (1979b): Über die Frostgefährdung von Fichten in verschiedener Höhenlage. Wiss. Mitt. Meteorol. Inst. Univ. München 35, 51-57

TRANQUILLINI, W. (1982): Frost-drought and its ecological significance. In: Lange, O. L., Nobel, P. S.; Osmond, C. B. & Ziegler, H. (eds.): Encyclopedia of Plant Physiology, New Series 12 B, Physiological Plant Ecology II. Springer, Berlin, 379-400

TRANQUILLINI, W. & HAVRANEK, W. M. (1985): Influence of temperature on photosynthesis in spruce provenances from different altitudes. Eidg. Anst. Forstl. Versuchswes. Ber. 270, 41-51

TRANQUILLINI, W. & PLANK, A. (1989): Ökophysiologische Untersuchungen an Rotbuchen (*Fagus sylvatica* L.) in verschiedenen Höhenlagen Nord- und Südtirols. Cbl. f. d. ges. Forstwesen 106, 225-246

TRANQUILLINI, W. & TURNER, H. (1961): Untersuchungen über die Pflanzentemperaturen in der subalpinen Stufe mit besonderer Berücksichtigung der Nadeltemperaturen der Zirbe. Mitt. Forstlichen Bundesversuchsanstalt Mariabrunn 59, 127-151

Address of the author:

Prof. Dr. W. Tranquillini, Institut für Botanik, Universität Innsbruck, Sternwartestraße 15, A-6020 Innsbruck, Austria

Man-induced changes at the alpine timberline of the Val Fenga (Silvretta, Switzerland) and their reflections in pollen diagrams (preliminary report)

Joachim Hüppe & Richard Pott

Summary

Palynological records of a subalpine/alpine mire in the upper Val Fenga (Fimbertal, Silvretta) are presented for the period since about 6000 yr B.P. (conv. ^{14}C data). Indicators for human impact on the natural environment are reflected in the pollen diagram by anthropogenic indicators (Cerealia pollen, NAP pollen especially of Gramineae and Cyperaceae and a lot of herbs and shrub species). The first indicators of man-made disturbances start at about 2000 yr B.P., and in the following periods disturbance phases are reflected in timberline oscillations with antagonistic pollen spectra of *Pinus* (e.g. *P. mugo, P. cembra*), *Picea abies* and the NAP pollen ratio of Gramineae, Cyperaceae, Lycopodium, Saxifragaceae and a lot of alpine/subalpine herbs. The pollen diagram of a subalpine *Drepanocladus* peat profile in an altitude of 2250 m is interpreted with reference to modern pollen rain values. Emphasis is placed on the indicative value of distinct plant communities with specific subassociations of the Aveno-Nardetum community.

Zusammenfassung

Die obere Waldgrenze im Val Fenga (Fimbertal, Silvretta), die derzeit im wesentlichen von Koniferen der Gattungen *Pinus, Picea* und *Larix* gebildet wird, ist als etwa 200 m breiter Übergangssaum vor allem das Ergebnis anthropo-zoogener Einflüsse. In diesem Übergangsbereich vollzieht sich mit Annäherung an die Höhengrenze des Baumwuchses eine zunehmende Auflockerung des Waldes in meist scharf umrissene Baumgruppen und einzelne Bäume, die mit einer Krummholzstufe aus *Pinus mugo*, *Alnus viridis* und Rhododendron-Gesellschaften an die alpine Vegetation grenzen. Durch pollenanalytische Untersuchungen werden anthropogene Eingriffe und entsprechende Waldgrenzschwankungen seit dem Beginn der wirtschaftlichen Nutzung (vor allem mit Alpweide) sichtbar gemacht. Durch Auswertung des aktuellen differenzierten Pollenniederschlags in distinkten Gesellschaftsausbildungen des Aveno-Nardetum im Umfeld eines alpin/subalpinen Moores wird ein erster Versuch spezieller "Pollenpräsentationstypen" im Sinne von POTT & HÜPPE (1991) versucht.

1. Introduction

Vegetation belts in high mountains are influenced by exposition, slope (inclination), and other local climatic or edaphic factors and cannot exactly be set out on the terrain. For example, the lower boundary of the subalpine belt is difficult to determine exactly just in the same way as the lower boundary of the alpine belt is to be distinguished from the area where the nival belt begins.

The natural conditions have been obscured in many places by human activities which covered areas e.g. of alpine meadows, for several hundred meters down into the valleys. However, under natural conditions grasslands without trees are found stretching from the alpine down to the montane belt. The woodland and tree lines as lower limits of the alpine regions are generally determined by ELLENBERG (1963, 1988), HOLTMEIER (1974, 1986, 1989) and KÖSTLER & MAYER (1970), and they are distinguished as a "forest line" (or woodland limit) where a more or less closed stand of many individual trees comes to a stop. A tree line (or tree limit) is determined by the level up to which isolated trees are able to climb. A line joining their highest outposts is called the cripple limit (Photo 1). These limits, especially those of the woodland and individual trees, are set either by nature or by man and his livestock. So it is only through man's influence that the woodland limit and the tree limit have been separated. In the Alps a few sharply defined closed woodland limits can still be found, even on slopes of deep soil which are not too steep, e.g. in the Silvretta area and in the neighboured Upper Engadin, as HOLTMEIER (1967) stated.

The woodland boundary in Graubünden and Tyrol, for example, has been lowered by human activities and livestock. However, as KRAL (1973) has shown for the Eastern Alps not only man and animals but also changes in the general climate can effect the altitude of the woodland limit. The actual woodland boundary is about 230 m below the potential one as determined by climate. Today, in the subalpine zone of the Fimbertal we can find some principal scrub or dwarf shrub communities, characterized by the dominance of *Pinus mugo* (Photo 2) or *Alnus viridis* (Photo 3). They are related to the more extensive and characteristic subalpine zone as a result of grazing, summer farming and other human impact. In the same way the Rhododendro-Vaccinietum heaths with Rhododendron species, which extend between the woodland and tree limits in the outer Alps and in some parts of the central Alps, are considered to indicate areas which have been woodland in former times.

Many investigators have worked on the problem of natural causes of the altitude limits of woodlands and of distinct tree species. More recently ecophysiologists in particular have taken up the challenge (e.g. BILLINGS, 1969; TRANQUILLINI, 1979; HADLEY & SMITH, 1986). They are all aware that no one single factor would be responsible for this problem but the whole character of the climate must be considered and the specific adaptation of the trees is of great importance. Discussions on a historical aspect concerning the development of the natural environment at the alpine timberline by palynology are quite rare (BORTENSCHLAGER, 1990).

Photo 1 Alpine tree limit and altitudinal sequence in the pattern of plant formations in the Val Fenga with Larici-Cembretum, alpine/subalpine grasslands, Junipero-Arctostaphyletum, and *Alnetum viridis*. Trees (*Larix europaea* and *Pinus cembra*) are able to survive on the ridges, where they are not disturbed by the cattle

Photo 2 Subalpine prostrate mountain pinewoods (krummholz-elfin wood) with *Pinus mugo* on lavinar slopes. Where the slopes and knolls consist of lime deficient rock but are nevertheless dry, *Pinus mugo* dominates

Photo 3 Subalpine green alder scrub with *Alnus viridis*. *Alnus viridis* prefers sites where water is readily available on acid soils

Photo 4 Scheuchzerio-Caricetea small sedge fen and spring swamps with *Eriophorum scheuchzeri*. The peat formations of this mire were formed during the Postglacial warm period and have thus to be considered as subfossil

Photo 5 Acid soil grassland of the lower alpine and the subalpine belts with *Nardus stricta*, *Diphasium alpinum*, and *Carex curvula* dwarf shrub

2. The timberline at Val Fenga (Fimbertal, Silvretta)

In Val Fenga we found a mire with 2.70 m organic layer of *Drepanocladus* peat in an altitude of 2250 m. The subalpine-alpine mire with an actual vegetation of dominant *Carex nigra, Eriophorum latifolium, Triglochin palustre, Drepanocladus revolvens, Drepanocladus exannulatus* and *Cratoneuron filicinum* (Photo 4 and Fig. 2) has been analysed. Because of the shallowness of the available sediments, samples were taken every 1 cm, and a standard laboratory preparation of boiling in 10% NaOH followed by acetolysis and, where necessary, cold HF was employed. Pollen was counted to a sum of between 200 and 500 (excluding spores and aquatics).

The preliminary results of the pollen analysis are presented in Fig. 1 (pollen diagram). Of course, our investigations are in the preliminary stages so that the picture shows only the general trend as known at the present time. As in all the mountainous sites men have been present well above the tree line, and the tree line today is considerably lower than the natural tree line would be under present climatic conditions.

It is possible to distinguish between a pine vegetation with high percentages of *Pinus* and increased amounts of Cyperaceae. At this initial phase of the development of the bog, *Picea* was also present. The migration and spread of *Picea* into the subalpine belt occured about 6000-5000 yr B.P. It is possible, and even observed in the Swiss Alps by WEGMÜLLER (1966) and LOTTER (1988), that during this migration the timberline rose and the vegetation of the subalpine region was characterized by *Picea* forests and on exposed places by *Pinus mugo* stands.

The influence of cultivation and the following re-occurance of the forest can be demonstrated very well in the profile of the Val Fenga (Fimbertal). The increasing pollen ratio of Cerealia in this spectrum as anthropogenic indicators and the correlation of increasing heliophytic shrubs, like *Salix, Corylus, Juniperus* and the increasing of the *Alnus* pollen ratio shows the same effects. It may be the time of prehistoric human influence and a first step of oscillation of the alpine timberline in this area by forest clearings for the creation and the establishing of pastures.

Special attention is drawn to the fluctuations in the NAP/AP pollen ratio. The phenomenon shows fluctuations of the former timberline in the Val Fenga too. In later periods we can see high amounts of Cyperaceae and a lot of alpine and subalpine shrubs and herbs. The decreasing effects of *Alnus* with a second rise of *Picea abies* may be caused by the Medieval and historical human activities, e.g. reforestation in the lower parts of the Fimbertal or grazing effects of summer farming. The traces of human influence are detectable in the pollen diagrams.

POLLEN DIAGRAM UPPER FIMBERTAL, SILVRETTA, 2250 m a.s.l., HEIDELBERGER HÜTTE

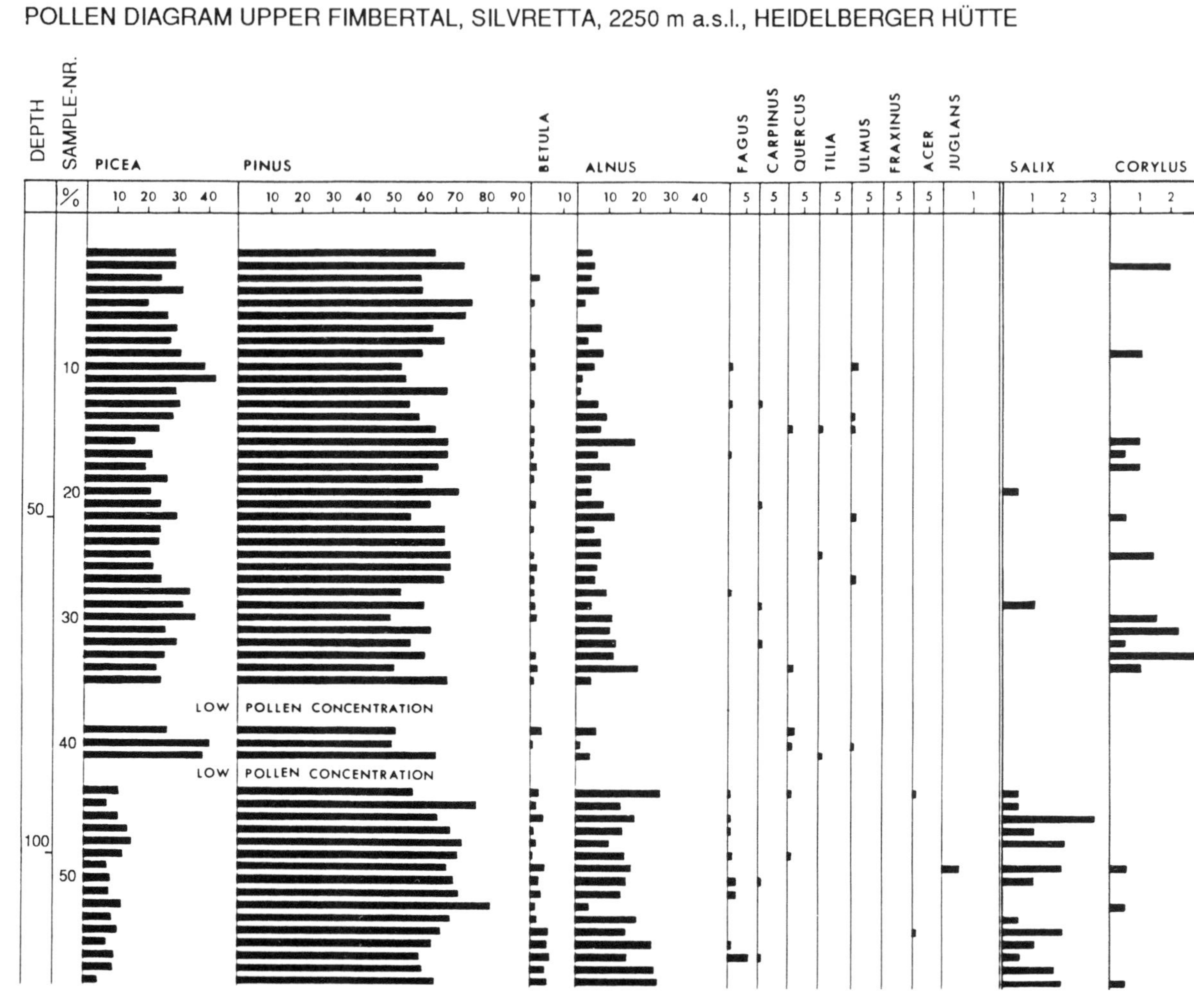

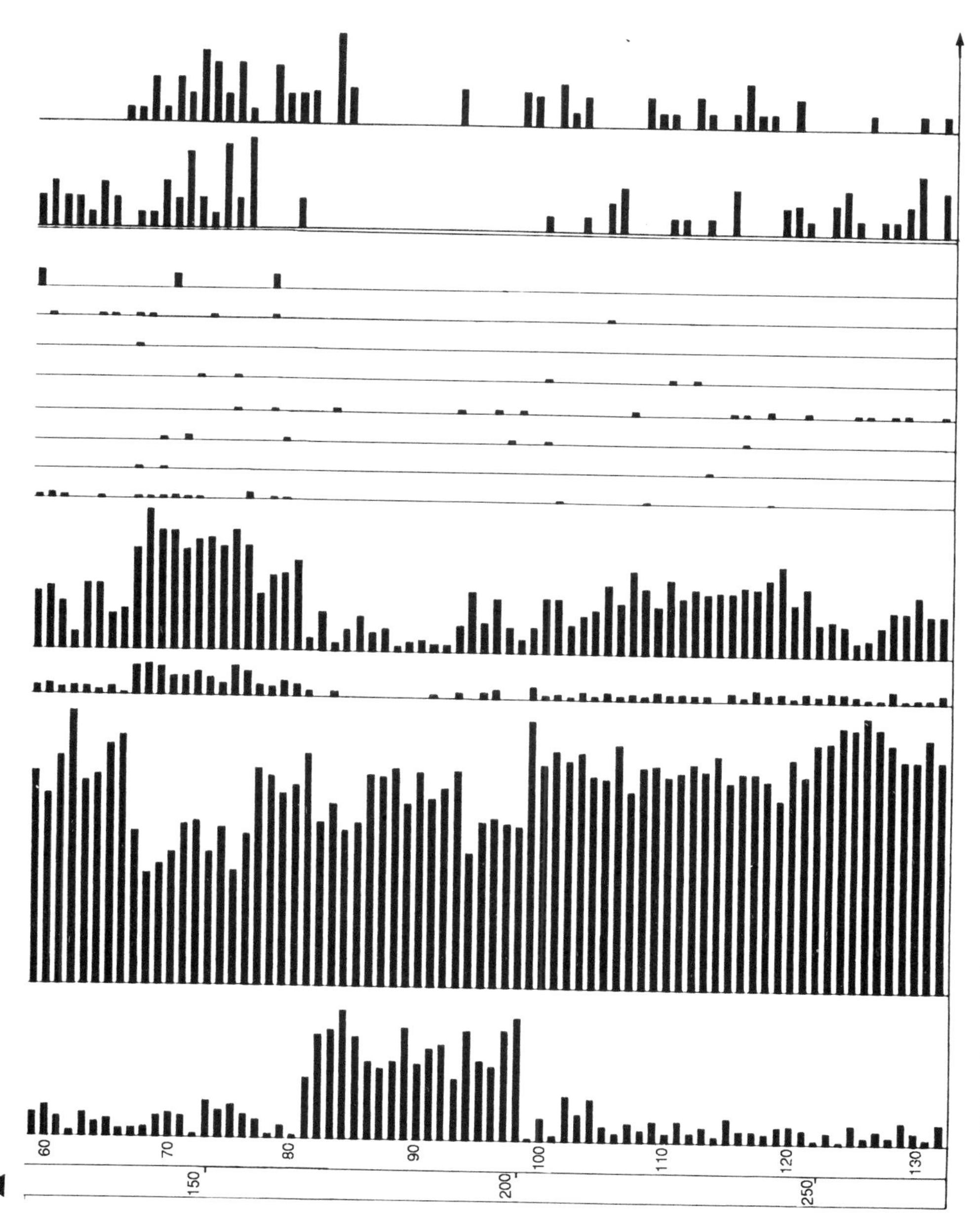

Fig. 1 Pollen diagram of the Fimbertal/Silvretta

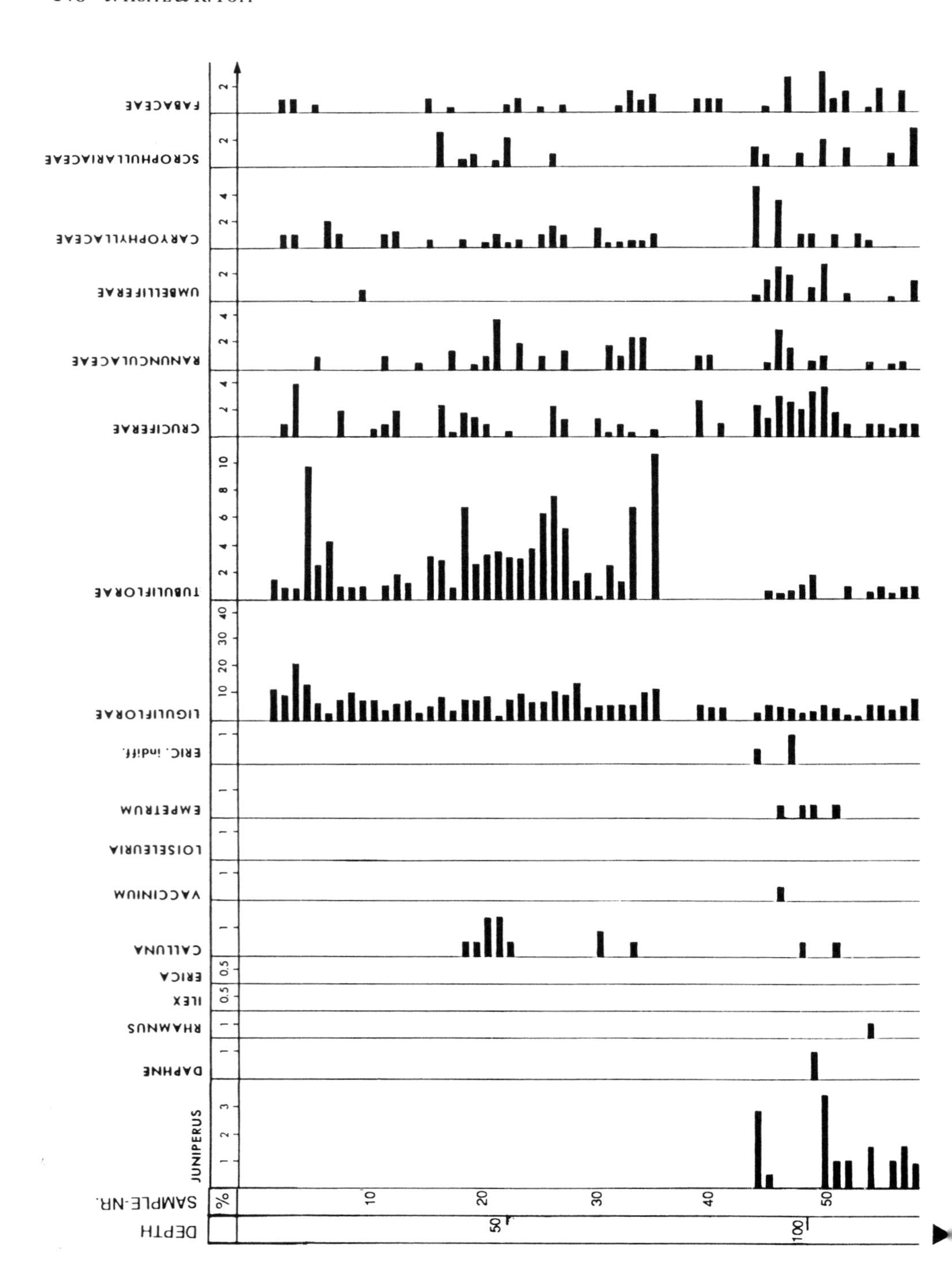
FABACEAE
SCROPHULARIACEAE
CARYOPHYLLACEAE
UMBELLIFERAE
RANUNCULACEAE
CRUCIFERAE
TUBULIFLORAE
LIGULIFLORAE
ERIC. indiff.
EMPETRUM
LOISELEURIA
VACCINIUM
CALLUNA
ERICA
ILEX
RHAMNUS
DAPHNE
JUNIPERUS
SAMPLE-NR.
DEPTH
%
10
20
30
40
50
50
100

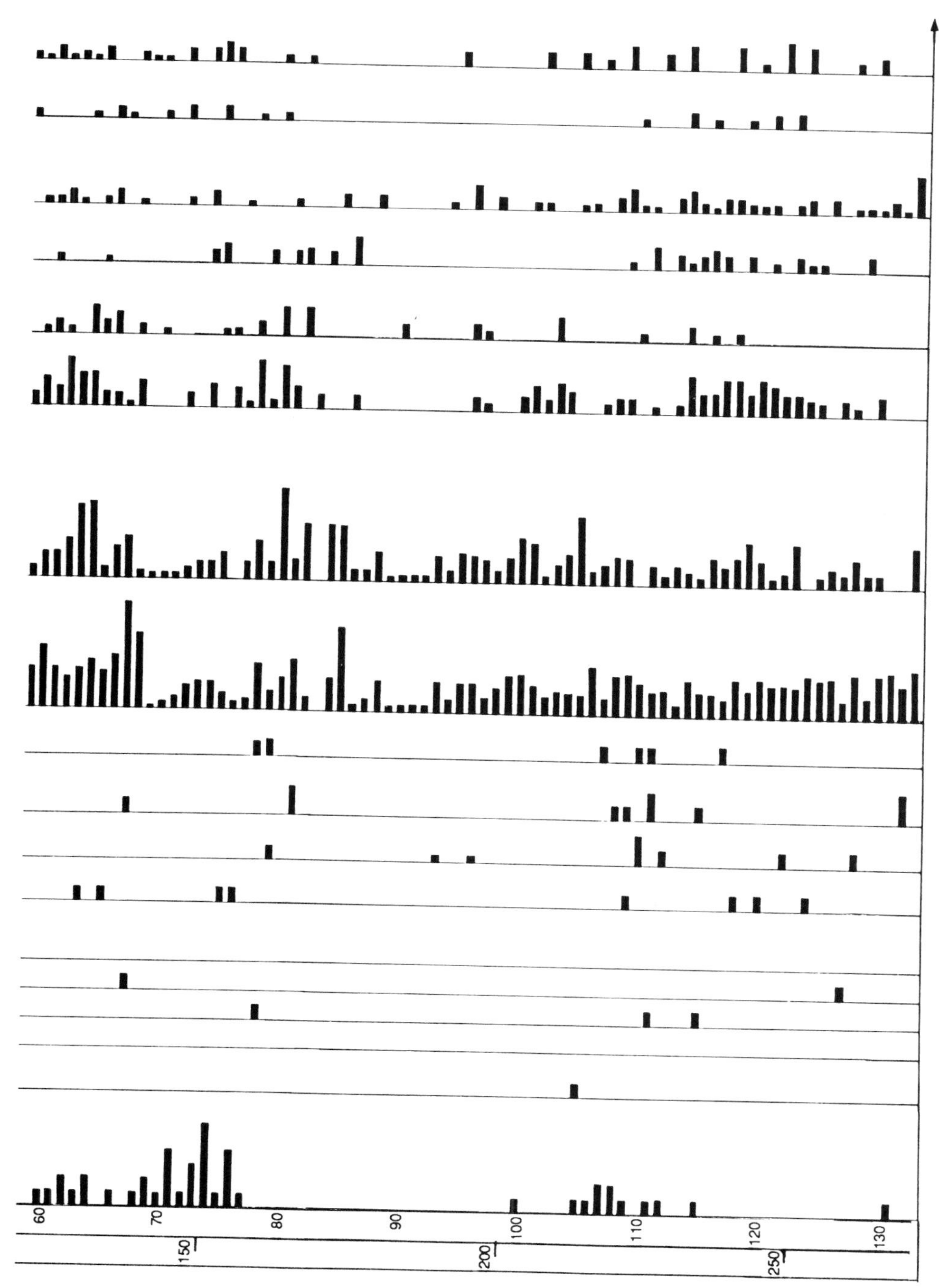

Fig. 1 Pollen diagram of the Fimbertal/Silvretta (continued)

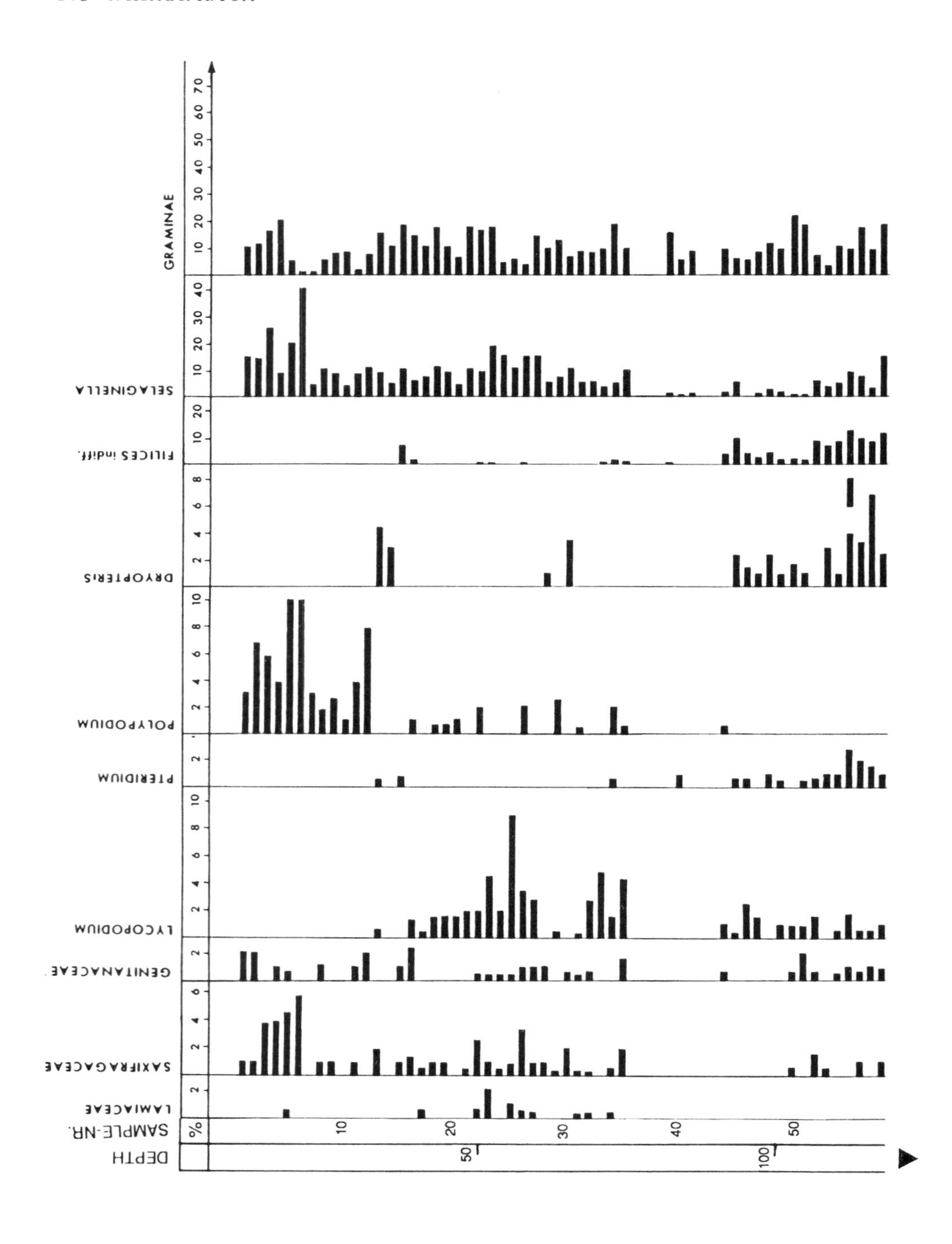
GRAMINAE
SELAGINELLA
FILICES indiff.
DRYOPTERIS
POLYPODIUM
PTERIDIUM
LYCOPODIUM
GENTIANACEAE
SAXIFRAGACEAE
LAMIACEAE
%
SAMPLE-NR.
DEPTH
10
20
30
40
50
100

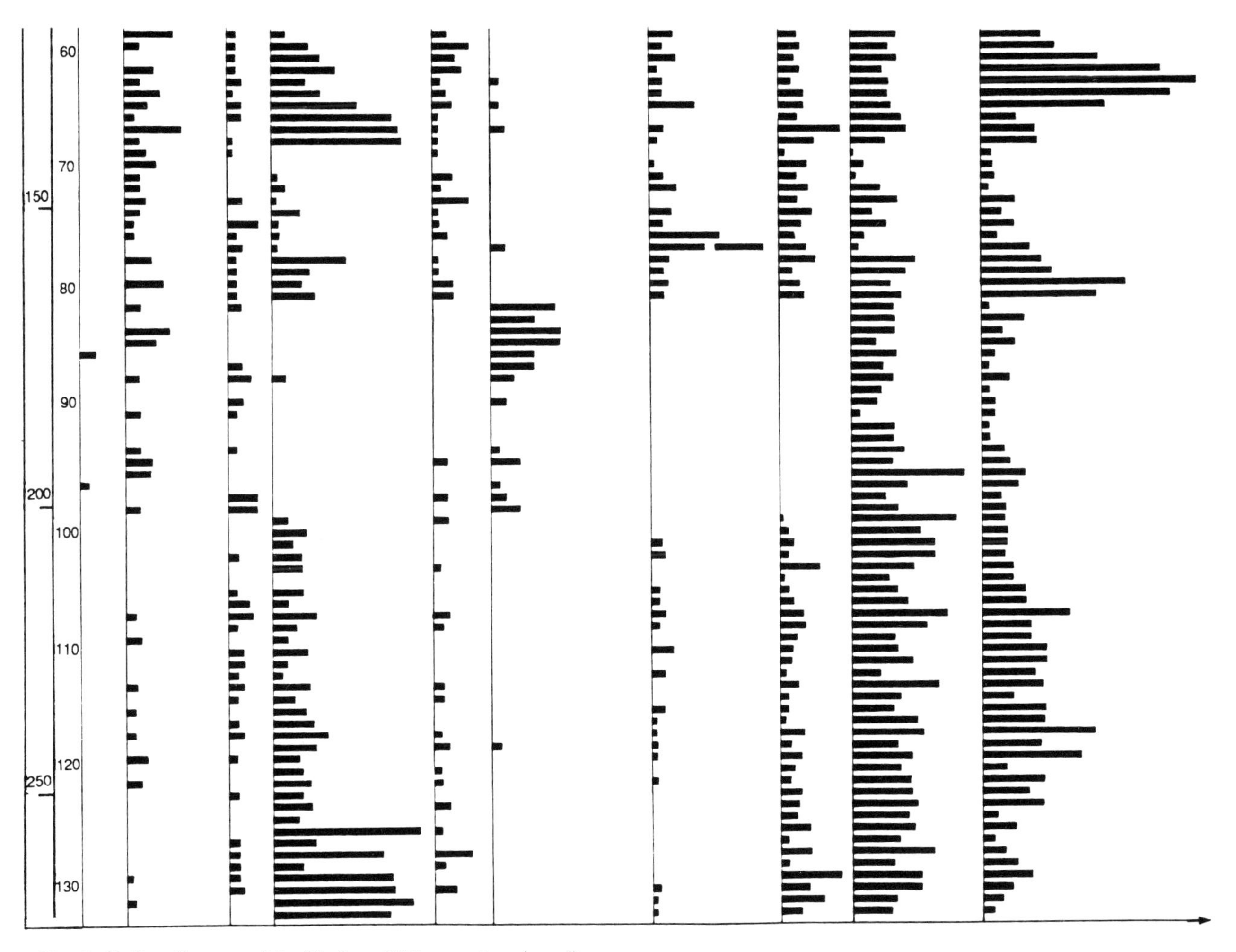

Fig. 1 Pollen diagram of the Fimbertal/Silvretta (continued)

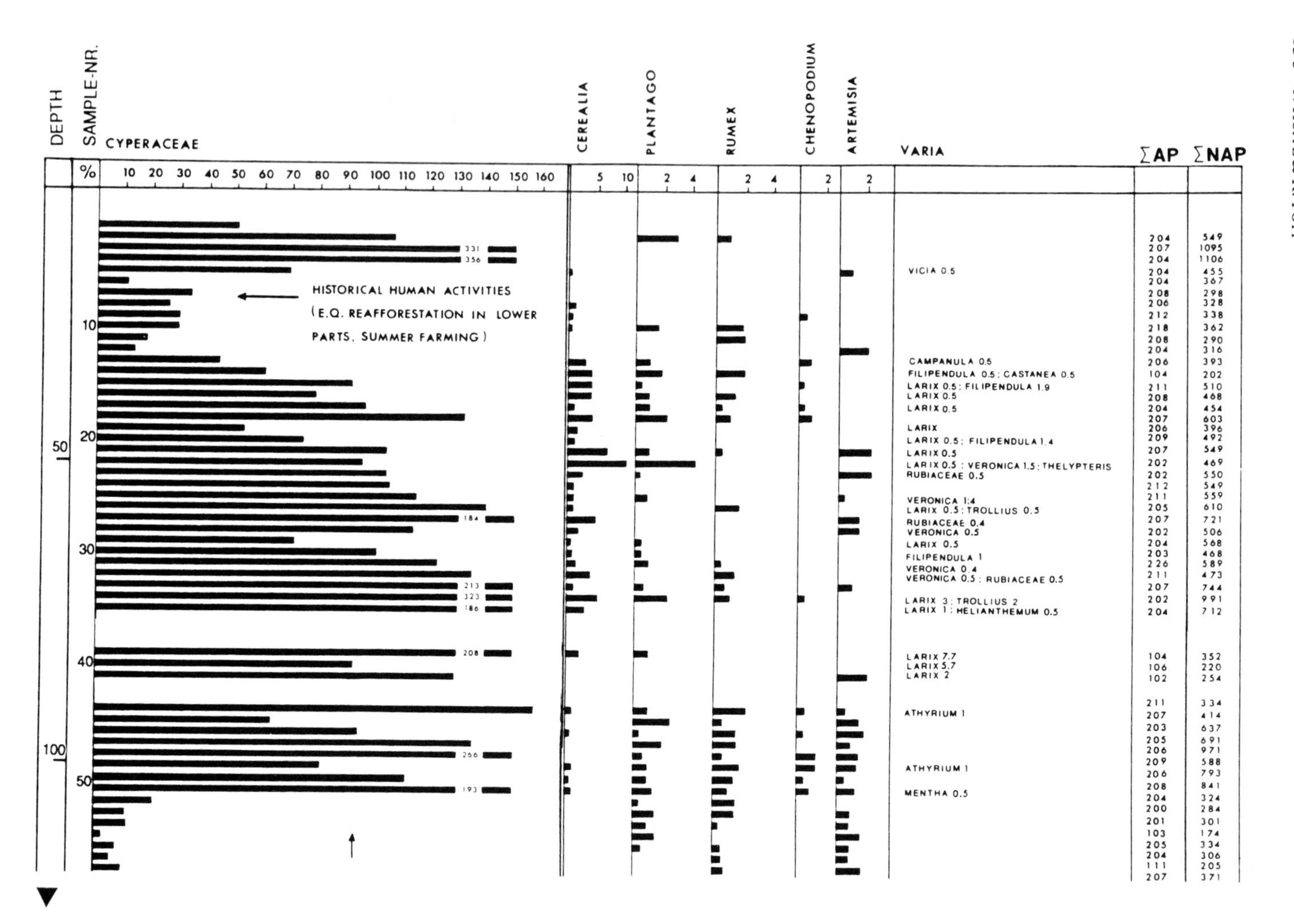

DEPTH
SAMPLE-NR.
%
CYPERACEAE
CEREALIA
PLANTAGO
RUMEX
CHENOPODIUM
ARTEMISIA
VARIA
∑AP
∑NAP
HISTORICAL HUMAN ACTIVITIES
(E.Q. REAFFORESTATION IN LOWER
PARTS, SUMMER FARMING)
VICIA 0.5
CAMPANULA 0.5
FILIPENDULA 0.5; CASTANEA 0.5
LARIX 0.5; FILIPENDULA 1.9
LARIX 0.5
LARIX 0.5
LARIX
LARIX 0.5; FILIPENDULA 1.4
LARIX 0.5
LARIX 0.5 : VERONICA 1.5; THELYPTERIS
RUBIACEAE 0.5
VERONICA 1.4
LARIX 0.5; TROLLIUS 0.5
RUBIACEAE 0.4
VERONICA 0.5
LARIX 0.5
FILIPENDULA 1
VERONICA 0.4
VERONICA 0.5 ; RUBIACEAE 0.5
LARIX 3; TROLLIUS 2
LARIX 1; HELIANTHEMUM 0.5
LARIX 7.7
LARIX 5.7
LARIX 2
ATHYRIUM 1
ATHYRIUM 1
MENTHA 0.5

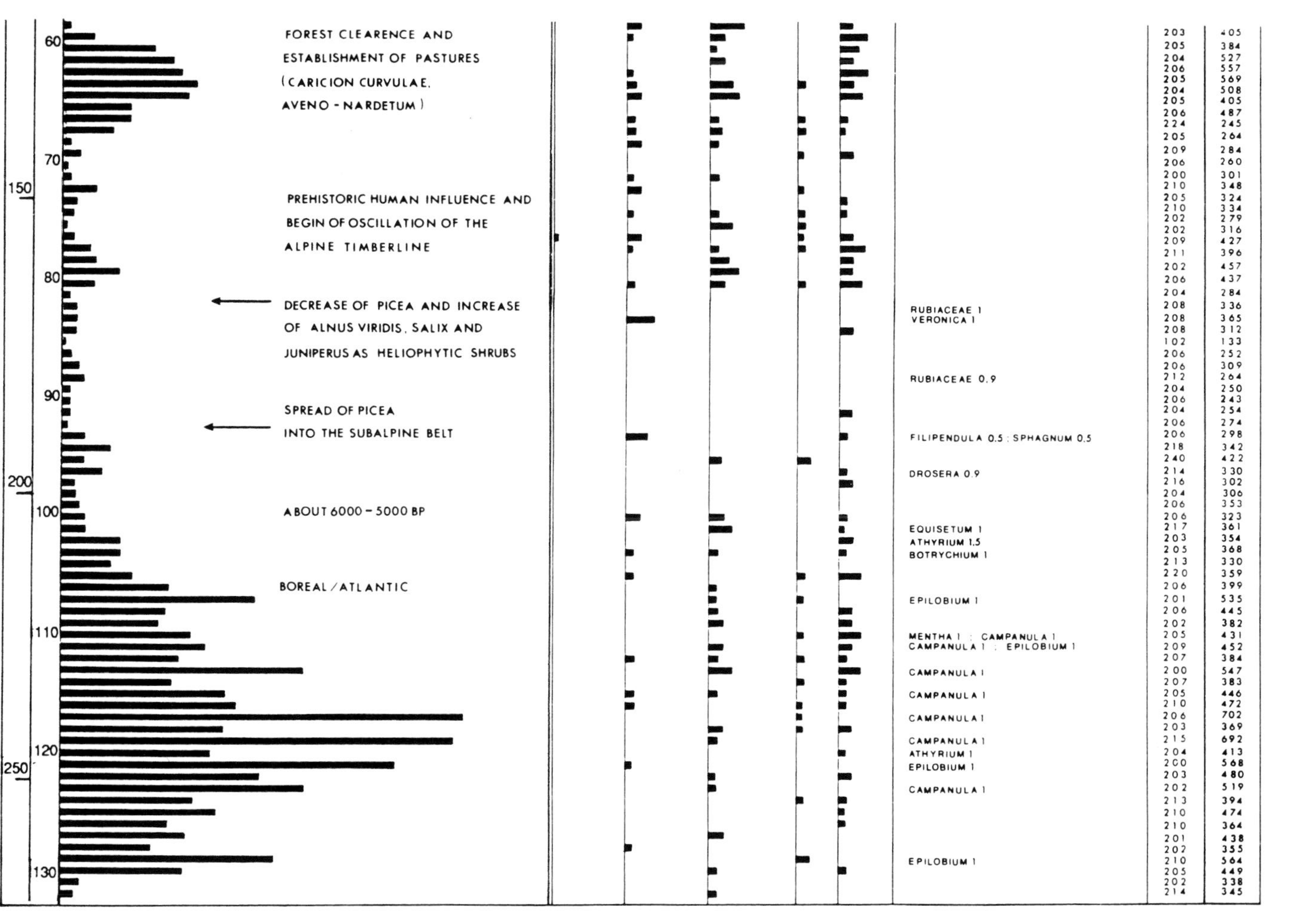

Fig. 1 Pollen diagram of the Fimbertal/Silvretta (continued)

3. The actual vegetation as a source of pollen deposition

The investigated alpine mire is situated about 200 m above the actual timberline. The dominant species of the timberline today are *Picea abies*, *Larix decidua* and some trees of *Pinus cembra* (Photo 1). The area between the closed forest line and the treeline is dominated by Erico-Mugetum with *Pinus mugo* and Junipero-Arctostaphyletum with *Juniperus nana* and *Arctostaphylos uva-ursi* (Photos 2 and 3).

The landscape in the neighbourhood of the mire is dominated by some rust-red Alpenrose heath community (Rhododendro-Vaccinietum), a lot of creeping *Azalea* carpets as espalier heaths (Loiseleurio-Cetrarietum) and the alpine Caricion curvulae vegetation on acid soils. A further grassland community surrounding the locality is the Aveno versicoloris-Nardetum coming up in an *Arnica montana* form and a *Carex curvula* form. Because *Nardus* is indirectly favoured by grazing, it is able to penetrate other communities in the subalpine and alpine belt (vegetation Table 1).

The actual vegetation surrounding the investigated mire at present (Fig. 2) covers the whole area. It consists of a relatively narrow sandy ridge with a present vegetation cover of Aveno-Nardetum. The regional vegetation cover is closely related to soil type with *Pinus mugo* forest on the moraine areas in lower altitudes and *Alnus viridis* or successional birch woodland on the more silty substrates. The flora, as a whole, is a relatively species-rich one, which is found on the adjacent oligotrophic and minerotrophic mainland.

4. Local and regional pollen deposition

The aim of our study is to establish modern pollen and vegetation relationships at the local scale and hence to provide a basis for interpreting fossil pollen assemblages in terms of past and actual vegetation.

The approach is an attempt to characterize a range of modern vegetation types on phytosociological base by means of contemporary pollen spectra and then to compare the fossil pollen assemblages with modern pollen spectra. Possible proportions between the modern and fossil spectra may be concluded as a similar distinct vegetation type in order to reconstruct past vegetation and vegetation changes in the alpine/subalpine environments.

Therefore surface samples (moss and lichen polsters or upper soil layers) from carefully chosen areas of contrasting vegetation around the mire locality (see Table 1) have been analysed palynologically. Particular attention was given to the identification at the lowest taxonomic level possible in our preliminary stage of investigation. Vegetational data have been collected for each sample site too (see Photo 5).

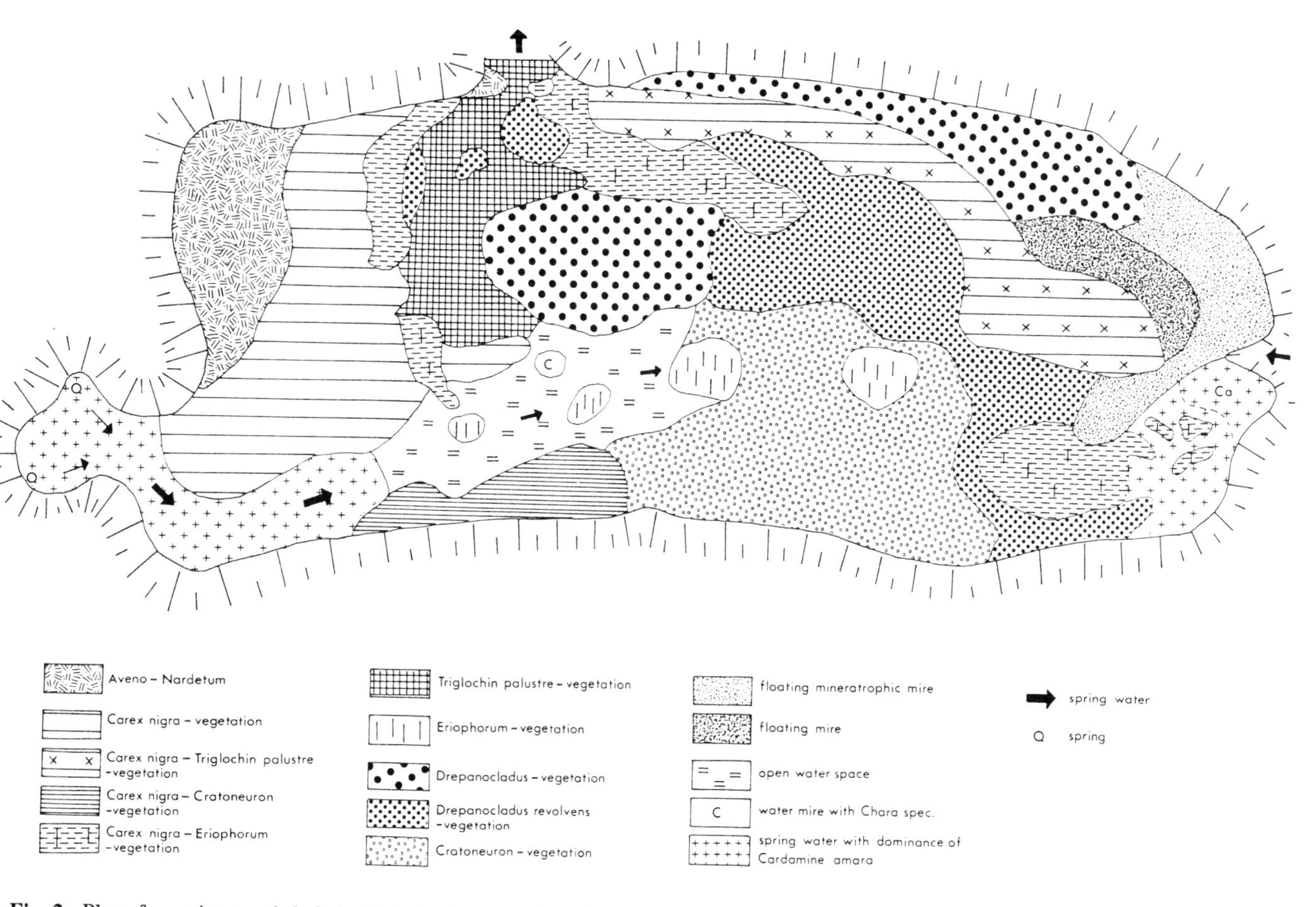

Fig. 2 Plant formations and their habitats in the alpine/subalpine mire, showing the various plant communities depending on the water and base contact of the organic material

Table 1 Phytosociological situation around the sampling sites (Val Fenga, 8.9.1989)

Aveno versicoloris-Nardetum Oberd. 1957
Sample No 1-2: *Arnica montana* form,
Sample No 3-5: *Carex curvula* form (=Curvulo-Nardetum Oberd. 1959)

Sample No	**1**	**2**	**3**	**4**	**5**
area (m^2)	10	16	6	12	4
vegetation cover (%)	100	95	75	100	70
altitude m above sea level	2200	2200	2237	2250	2400
VC Nardion					
Nardus stricta	4	4	2	2	1
Potentilla aurea	1	1	+	2	2
Gentiana acaulis		+	+	+	+
Leontodon helveticus	2			1	2
Alchemilla alpina				+	
Pulsatilla apiifolia	+		+		+
Solidago virgaurea ssp. minuta		+			
Diff. Aveno-Nardetum (Caricion curvulae-species)					
Avena versicolor	2	2	1	2	1
Agrostis rupestris	+	1	2	1	2
Phyteuma hemisphaericum	1	1	+	1	1
Veronica bellidioides	1	+	1	+	1
Senecio carniolicus	1				1
Arnica montana-form					
Arnica montana	1	1	+		
Diphasium alpinum	1	1			+
Campanula barbata	2	1			
Luzula multiflora	1	1			
Geum urbanum	1	+			
Calluna vulgaris		1	+		
Phyteuma betonicifolium	+	+			
Antennaria dioica		1			
Hieracium pilosella		1			
Trifolium alpinum	+				
Carex curvula-form					
Carex curvula		+	1	2	3
Euphrasia minima			1		+
Silene acaulis			+	1	
Antennaria carpatica			1		

Lichens					
Cetraria islandica	1	1	1	1	2
Cladonia rangiferina	1	1	+	1	1
Cornicularia aculeata	+	1	1	+	+
Cladonia impexa	r	r	+	+	+
Cetraria nivalis			1	1	2
Cladonia chlorophaea	+		+	+	+
Cladonia pyxidata			1		+
Cladonia foliacea				+	1
Cladonia furcata					1
Additional species					
Homogyne alpina	1	1		1	1
Euphrasia rostkoviana	1		1	1	1
Polygonum viviparum	1		1	1	1
Carex sempervirens	+	1	1	+	
Selaginella selaginoides	1	+		+	+
Chrysanthemum alpinum	1	+		1	
Arctostaphylos uva-ursi		+	+	+	
Loiseleuria procumbens			1		2
Hieracium villosum	1				1
Vaccinium vitis-idaea		1			1
Campanula scheuchzeri			1	1	
Vaccinium myrtillus	+				1
Ranunculus montanus		+			1
Galium anisophyllum			+	1	
Crepis aurea	+			+	
Sempervivum arachnoideum			+		+
Primula farinosa			+	+	
Poa alpina	1				
Silene cucubalus	1				
Avenella flexuosa		1			
Nigritella nigra			1		
Trifolium pratense sspnivale				1	
Vaccinium uliginosum				1	
Carex ferruginea		+			
Myosotis alpestris		+			
Anthoxanthum alpinum				+	
Carex ericetorum				+	
Carex ornithopoda				+	
Achillea millefolium				+	
Salix herbacea				+	
Mosses					
Polytrichum piliferum	2	2	1	1	2
Polytrichum juniperinum	1	1	+	1	1
Dicranum scoparium				1	1
Rhacomitrium canescens					+

The results show the strong gradients of vegetation cover and the phytosociological evidence influencing the observed patterns within the modern pollen aspects. The distribution of single species within a phytocoenosis is also significantly variable, and the relationship of certain pollen types to particular vegetation texture is evident. We think that it is possible to explore statistically in the future, which will provide a usuable instrument to detect the relationship of pollen and vegetation at a local as well as at a regional scale (e.g. HÜPPE et al., 1989).

For the interpretation of pollen diagrams the modern pollen rain is an important monitoring factor. In this way pollen percentage data show the possibility to make estimates

- of the degree and the type of pollen presentation;
- of man-induced situations in general;
- of distances from the sampling site;
- of abundances within concentric circles around the sampling site, and
- of the nature of the intervening and surrounding vegetation.

It will therefore be possible to demonstrate the different vegetation units within a pollen diagram and to clear up changes in vegetation patterns in time and space.

The combination of the interpretation of surface samples and pollen analysis of fossil peats within the same locality will be a method to detect and date the local or regional vegetation history on a palynological base and to clear up the development of local vegetation units. It is an attempt to describe the vegetation and landscape dynamics in time and space (POTT & HÜPPE, 1991).

5. Regional pollen deposition and local pollen sources

The detection of regional pollen, i.e. the pollen deposition away from the influence of local pollen sources, within the analysis of recent surface samples is an important step in the spatial reconstruction of the development of plant cover. In pollen analysis this concept is useful to detect the relation between plant formation, landscape and regional pollen deposition. In landscapes with large treeless areas like Val Fenga it is possible to determine

- the regional and extra regional pollen deposition out of the subalpine forest zone;
- the long distance transport (e.g. *Juglans, Fagus*), and
- the recognition of the local pollen deposition reflecting vegetational events on a fine scale.

Results of available surface samples are presented too (Fig. 3-5).

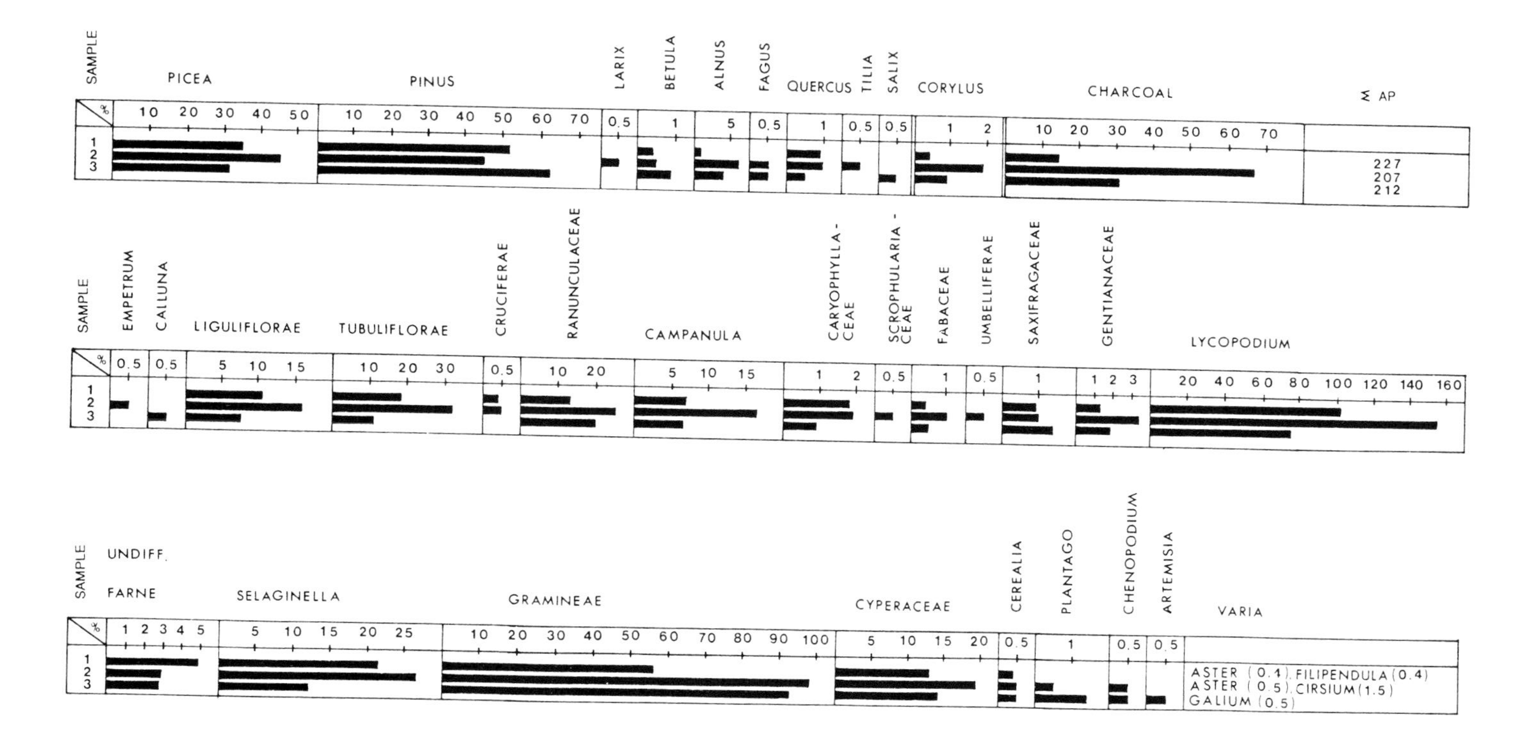

Fig. 3 Surface samples from Fimbertal (I): Aveno versicoloris - Nardetum, *Arnica montana* - Form (2200 m a.s.l.)

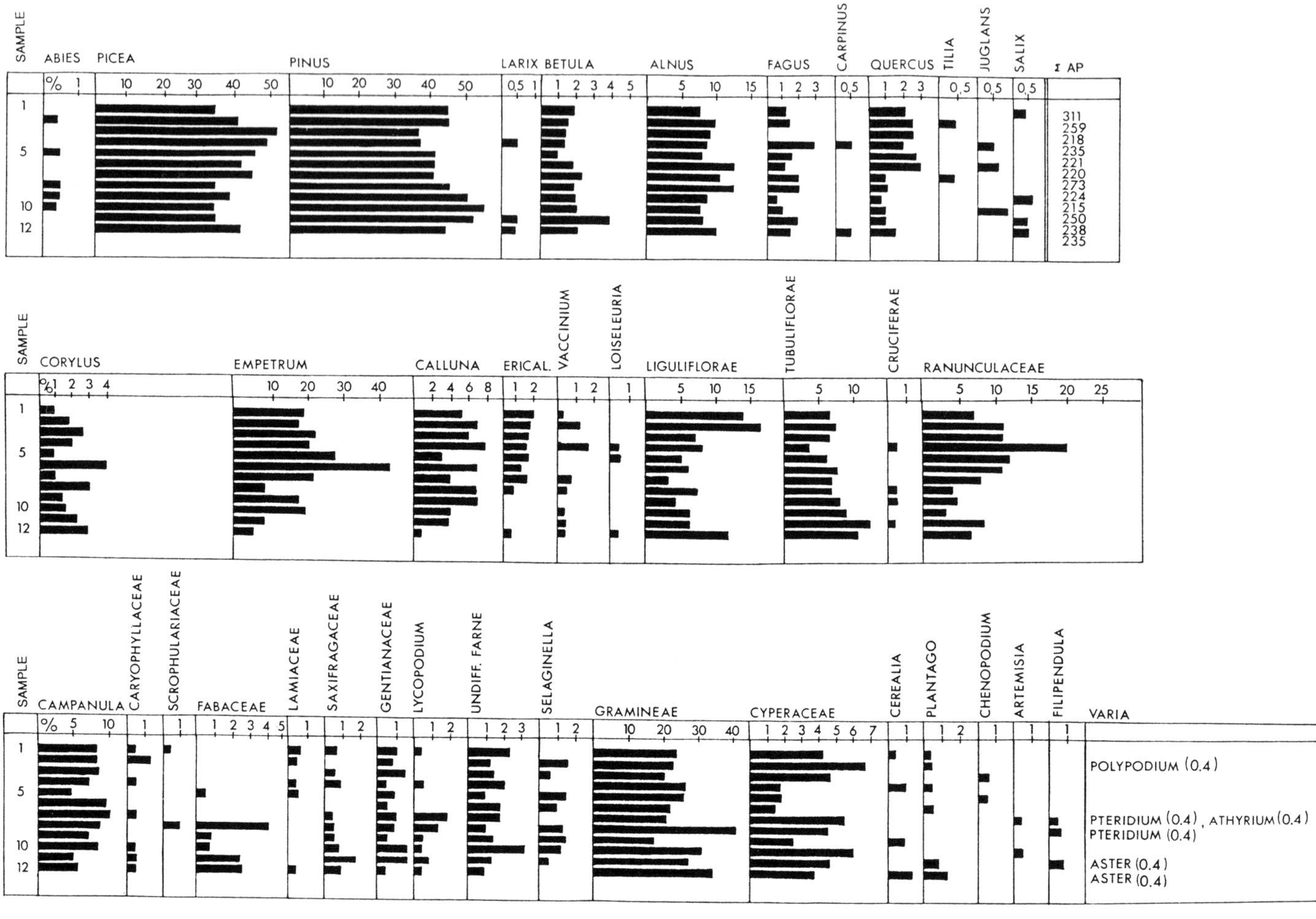

Fig. 4 Surface samples from Fimbertal (II): Aveno versicoloris - Nardetum, *Carex curvula* - Form (2237 m a.s.l.)

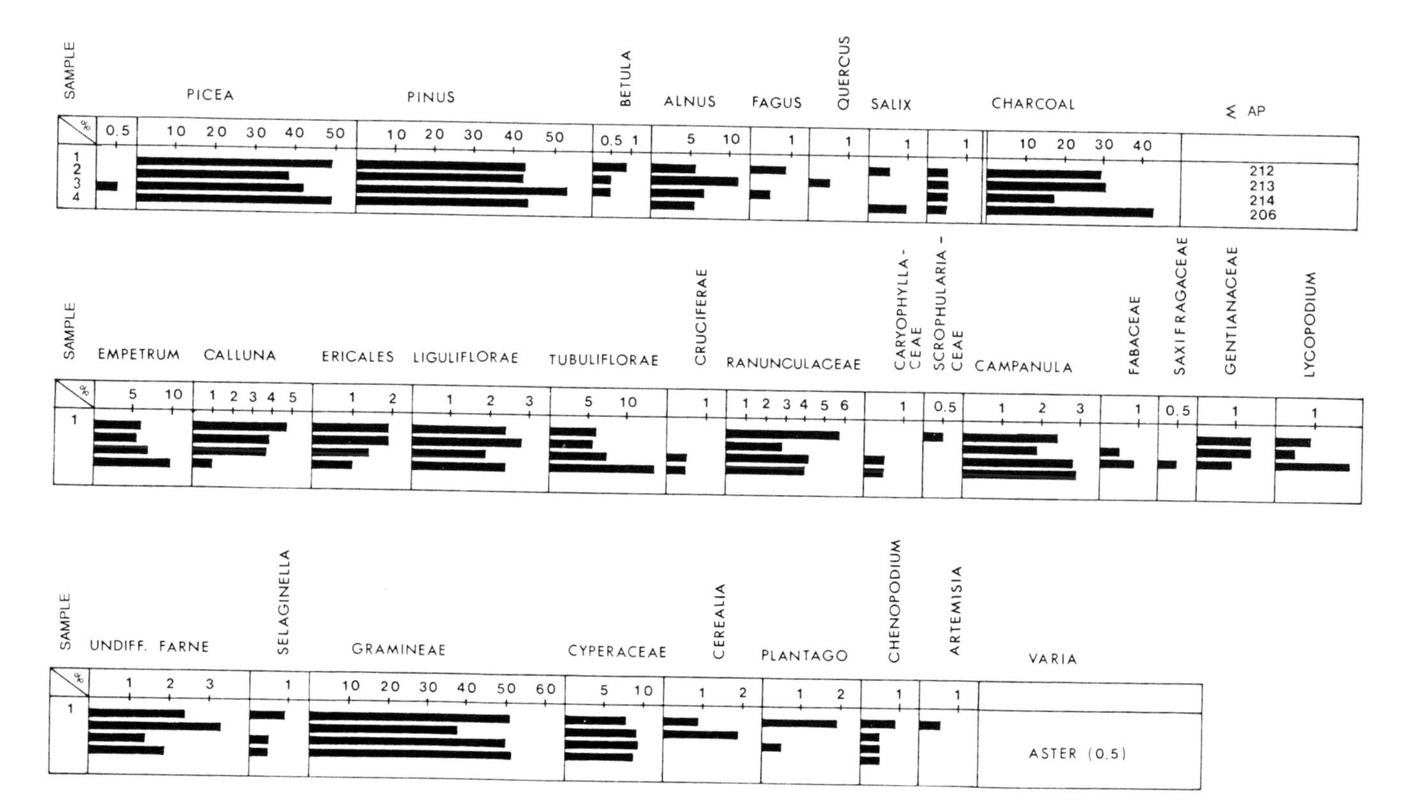

Fig. 5 Surface samples from Fimbertal (III): Aveno versicoloris - Nardetum, *Carex curvula* - Form (2400 m a.s.l.)

The surface samples are grouped according to the vegetation type of Aveno-Nardetum around the mire and they show quite strong relations between the percentage representation of the pollen types for each different vegetation type. It can be demonstrated (Fig. 3-5) that the difference in pollen influx between these distinct vegetation types may vary a little bit, but the percentage representation remains constant. For each group of plants the relevant distinguishable pollen types are listed (see Fig. 3-5, pollen diagrams). Wherever pollen identification is only possible to the level of the family or even genus, the value of that pollen type as an indicator is obviously diminished.

A closer appreciation of the relation between pollen reaching the area and the vegetation of the site has to be mentioned. The dense *Drepanocladus* peat filtrates and conserves the majority of the local and regional pollen of the open alpine fields, the presentation of forest species of the subalpine timberline is represented too. The openness of the field situatuion allows the long-distance dispersal of pollen over a greater distance.

In the future this kind of research strategy and information about palynological data will be necessary to improve our knowledge of changes of the timberline, induced by man or by natural factors. In this special area of Val Fenga (Fimbertal, Silvretta) we can get a lot of information furnished by several other ombrotrophic mires between the recent and the former timberline positions. Further palynological investigations with overlapping series and radiocarbon dates will constitute a progress of palaeoclimatic reconstructions as well as in demonstrating changes in the floristic composition and extension of forest ecosystems at the alpine timberline.

References

BILLINGS, D.W. (1969): Vegetational pattern near alpine timberline as effected by fine-snow drift interactions. Vegetatio 19, 192-207

BORTENSCHLAGER, S. (1990): Beiträge zur Vegetationsgeschichte Tirols I-IV. Innsbruck, 130 p.

ELLENBERG, H. (1963): Vegetation Mitteleuropas. 1st ed., Ulmer, Stuttgart, 981 p.

ELLENBERG, H. (1988): Vegetation ecology of Central Europe. 4th ed., Cambridge, 731 p.

HADLEY, J. L. & SMITH, W. K. (1986): Wind effects on needles of timberline conifers: seasonal influence and mortality. Ecology 67/1, 12-19

HOLTMEIER, F. K. (1967): Zur natürlichen Wiederbewaldung aufgelassener Alpweiden im Oberengadin. Wetter u. Leben 19, 195-202

HOLTMEIER, F. K. (1974): Geoökologische Beobachtungen und Studien an der subarktischen und alpinen Waldgrenze in vergleichender Sicht (nördliches Fennoskandien/Zentralalpen). Erdwiss. Forsch. 8, 130 p.

HOLTMEIER, F. K. (1986): Die obere Waldgrenze unter dem Einfluß von Klima und Mensch. Abh. Westf. Mus. Naturk. 48, 2/3, 395-412

HOLTMEIER, F. K. (1989): Ökologie und Geographie der oberen Waldgrenze. In: Pott, R. (ed.): Berichte der Reinh.-Tüxen-Ges. 1, Hannover, 15-45

HÜPPE, J.; POTT, R. & STÖRMER, D. (1989): Landschaftsökologisch-vegetationsgeschichtliche Studien im Kiefernwuchsgebiet der nördlichen Senne. Abh. Westf. Mus. Naturk. 51/3, 77 p.

KÖSTLER, N. & MAYER, H. (1970): Waldgrenzen im Berchtesgadener Land. Jahrb. Ver. Schutz Alpenpfl. Tiere 35, 1-33

KRAL, F. (1973): Zur Waldgrenzdynamik im Dachsteingebiet. Jahrb. Ver. Schutze Alpenpfl. Tiere 38, 71-79

LOTTER, A. (1988): Paläoökologische und paläolimnologische Studie des Rotsees bei Luzern. Pollen-, großrest-, diatomeen- und sedimentanalytische Untersuchungen. Diss. Bot. 124, 187 p.

POTT, R. & HÜPPE, J. (1991): Die Hudelandschaften Nordwestdeutschlands. Abh. Westf. Mus. Naturk. , 313 p.

TRANQUILLINI, W. (1979): Physiological ecology of the alpine timberline. Ecol. Stud. 31, 1-131

WEGMÜLLER, S. (1966): Über die spät- und postglaziale Waldgeschichte des südwestlichen Jura. Beitr. Geobot. Landesaufn. Schweiz 48, 1-143

Addresses of the authors:

Dr. J. Hüppe, Universität Hannover, Institut für Geobotanik, Nienburger Str. 17, D-3000 Hannover 1
Prof. Dr. R. Pott, Universität Hannover, Institut für Geobotanik, Nienburger Str. 17, D-3000 Hannover 1

Pollen analytical evidence of Holocene climatic fluctuations in the European Central Alps

Conradin A. Burga

Summary

The main problems and approaches for reconstruction of the palaeoclimate with pollen analysis are discussed. The present state of palaeoclimatological knowledge is based mainly on qualitative information. More quantitative results on palaeotemperatures, palaeo-precipitations, timberline and glacier fluctuations, and on the magnitude of rates of change should be investigated. Local and regional studies, multi-disciplinary investigations and palynological data banks of the Alps can provide more detailed information on the various palaeoclimate patterns and their gradual changes.

Zusammenfassung

Die Hauptprobleme und Lösungsansätze der palynologischen Erforschung des Paläoklimas der Alpen werden kurz diskutiert. Der bisherige Forschungsstand beruht hauptsächlich noch auf qualitativen Informationen. Es sind vermehrt quantitative Ergebnisse zu den Paläotemperaturen, den -niederschlägen, den Schwankungen von Wald- und Schneegrenze, den Gletscherbewegungen und zu Größenordnungen von Änderungsraten anzustreben. Lokale und regionale Untersuchungen, interdisziplinäre Ansätze sowie palynologische Datenbanken können detailliertere Angeben zum Mosaik des Paläoklimas der Alpen liefern.

1. Problems in finding evidence of climatic fluctuations

We have to consider the following problems when we are looking for palynological evidence of climatic fluctuations:

- plant succession and climate change; soil development in the Alpine foreland related to the ice-melting phases;
- anthropogenic influences and climate fluctuations (mainly since 5000 yr B.P.); overlaps of natural and anthropogenic impacts of vegetation;
- problems of reconstruction of the vegetation in former times in general; vegetation belts in the high mountain area;

- non-climatic plant migrations between 12,000 and 4000 yr B.P., e.g. *Pinus sylvestris, Pinus mugo, Pinus cembra, Larix decidua, Picea abies, Abies alba, Alnus viridis, Fagus sylvatica*;
- effects of plant ecology competition, e.g. shade-bearing trees versus light-demanding trees;
- regional and local climate effects, e.g. precipitation, wind;
- the use of the time-scale; interpretation of climate impacts under consideration of different time-scales.

All these aspects and even more have to be considered for the palaeoclimatological interpretation of lake and peat-bog sediments.

2. Reconstruction of climate changes by pollen analysis

Suitably sensitive areas and altitudes in the Alps are used for palynological palaeoclimate research. Investigations within the area of Holocene timberline fluctuations generally provide good palaeoclimate results (Burga, 1988). First investigations in such sensitive areas were carried out at the beginning of this century. It was Hager (1916) who investigated Central Alpine peat-bogs and macrofossils close to the potential and actual timberline. Ancient peat forming phases during the Middle Holocene temperature optimum between 8000 and 5000 yr B.P. (cp. for example Burga, 1991) are known.

Explanation to Fig. 1

	Northern Hemisphere		Southern Hemisphere
1)	Budyko, 1990	1)	Heusser, 1974
2)	Röthlisberger, 1986	2)	Seltzer, 1990
3)	Imbrie & Palmer Imbrie, 1986	3)	Clapperton, 1990
4)	Porter & Orombelli, 1985	4)	Markgraf, 1980a
5)	Gribbin & Lamb, 1978	5)	Röthlisberger, 1986
6)	Kellogg, 1978	6)	Clapperton, 1990
7)	Ammann, 1982	7)	Clapperton, 1990
8)	v. Rudloff, 1980	8)	Suggate, 1990
9)	Gaillard, 1985	9)	Brown, 1990
10)	Birks, 1991	10)	Graf, 1981
11)	Rognon, 1983	11)	Kuhry, 1988
12)	Faniran & Jeje, 1983	12)	Markgraf, 1989
13)	Brun, 1979	13)	Bigarella & Ferreira, 1985
14)	Lézine & Casanova, 1989		
15)	Neftel, Oeschger, Staffelbach & Stauffer, 1988	16)	Kukla, 1978
17)	Isla, 1989	18)	Frankenberg, 1986

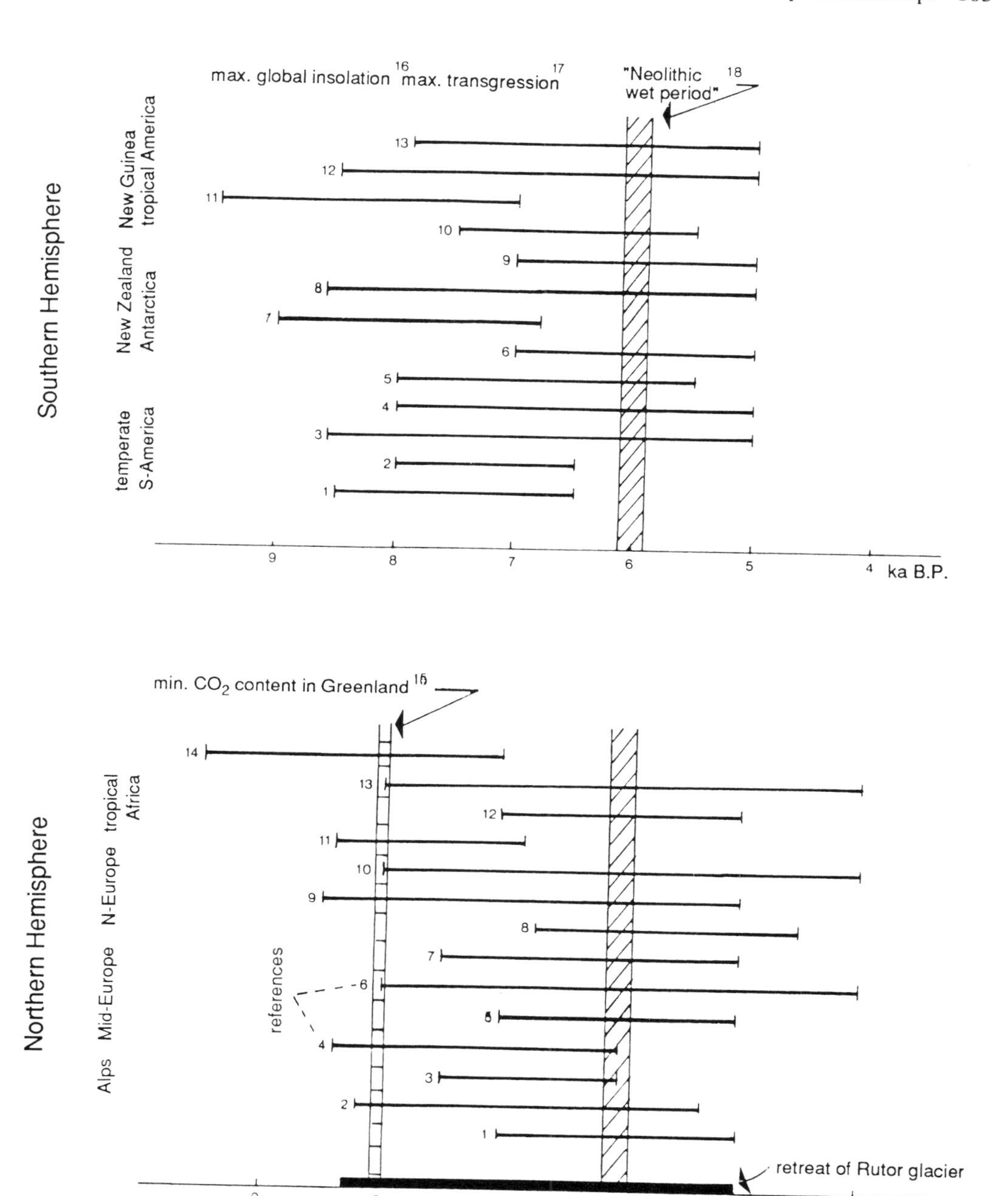

Fig. 1 Global evidence of the Postglacial Climate Optimum (numbers refer to references)

In several regions of the Central Alps different non-anthropogenic changes of vegetation structure during the Early and Middle Holocene can be stated, e.g. the recession of *Pinus cembra* in the Central Alps during the Boreal (Schamser-Oscillation, see BURGA, 1990). The reason for this recession is not known exactly, but it could be due to frost dryness as a result of decreasing winter precipitation.

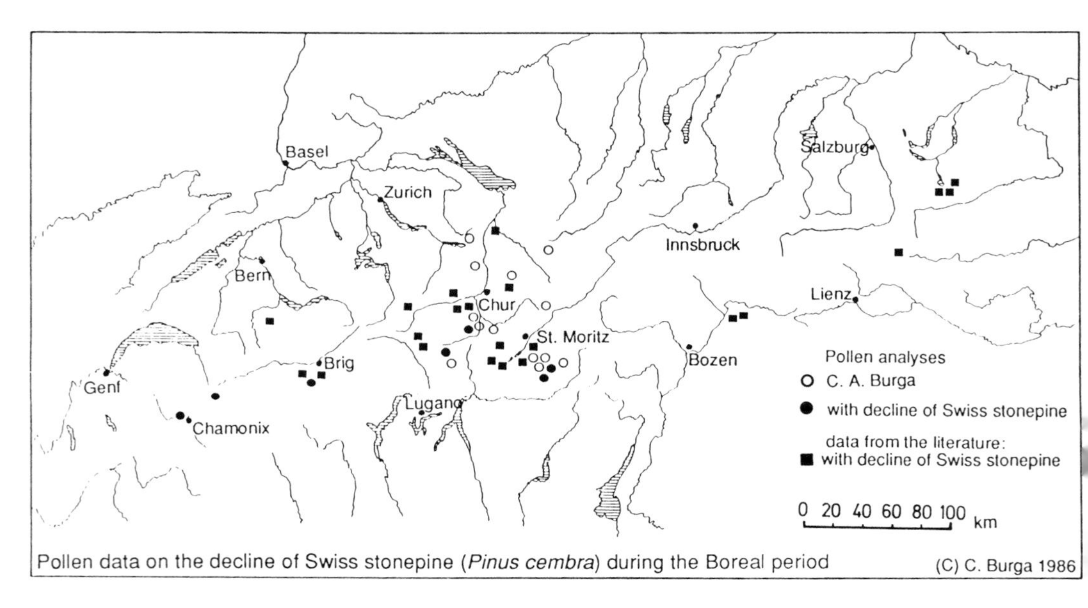

Fig. 2 Evidence to the *Pinus cembra* recession during the Boreal (symbols refer to recession sites)

Principles of climate reconstruction (temperature, precipitation) are based on the actualism, i.e. the palynological reconstruction of climate changes with the help of actual analogous vegetation types for which climate parameters are known (cp. the approaches by IVERSEN, 1944, and FRENZEL, 1967).

Generally two methods of climate reconstruction are used:

(1) the approach of IVERSEN (1944), GRICHUK (1964), and FRENZEL (1967). They used the coincidence of present-day distribution patterns of individual plant taxa with certain parameters of climate, like mean temperatures of the warmest or the coldest months;
(2) another approach is the plant geographical comparison between fossil and actual floras under consideration of the climate conditions for the occurrence of the plant taxa (cp. for example GRICHUK, 1984).

3. Pollen analytic evidence: present state of knowledge

Mainly qualitative results are available, e.g. from the Swiss Central Alps (ZOLLER, 1977; BURGA, 1979, 1987, 1988, 1991) and the French Alps (WEGMÜLLER, 1977; BOREL et al., 1984). A European overview was published by FRENZEL (1977) and GROVE (1979). Table 1 gives some palynological facts on the climate fluctuations Palü (Preboreal), Löbben (Subboreal) and Göschener cold phases 1 and 2 (Older Subatlantic). Quantitative results are, however, more desirable. There is little information on palaeotemperatures, provided by the reconstruction of timberline and glacier fluctuations (e.g. RÖTHLISBERGER, 1986; BURGA, 1991). Reconstructions of palaeoprecipitations are much less abundant. Qualitative results provide lake level fluctuations, peat-bog growth rates, and in some cases changes of floral elements due to change of moisture demand.

4. Future prospects

Palynological data banks provide in most cases a good basis for quantitative investigations. One data bank for Europe has been established by HUNTLEY & BIRKS (1983). The first palynological and palaeobotanical Quaternary data bank for the whole Swiss area, which provides full information about published material until the end of 1991, has been prepared by BURGA (publication in prep.). Research in the Swiss Alps must also continue with case studies of sensitive areas. The following two approaches should be considered:

(1) Studies of the actual local and regional pollen rain and the comparison with actual and fossil pollen floras (cp. AARIO, 1944; WELTEN, 1950; HEIM, 1971; JOCHIMSEN, 1972, 1986; TAUBER, 1977; TRIAT-LAVAL, 1978; DAMBLON, 1979; MARKGRAF, 1980; SCHNEIDER, 1984; BURGA, 1984);
(2) multi-disciplinary investigations of alpine and subalpine peat-bog, mire and lake sediments, and soil profiles. The investigation of palaeoenvironments in sensitive areas could provide more detailed information about palaeoclimate patterns.

Figures 3 and 4 show the relation between records of NAP phases at different altitudes and during different times. For the Swiss Central Alps and Prealps a decrease of evidence of NAP phases for the period between 10,000 and 1000 yr B.P. can be stated. Only areas located at a higher altitude give palaeoclimatological evidence for the Younger Holocene. It seems to reflect a decrease of intensity of climate change in contrast to the Late Glacial from the Early to Younger Holocene.

Concerning the lower located areas of the Jura Mountains and the Swiss Plateau we find only a few records of NAP phases which can be considered as climate changes. This result seems to underline the statement above. In any case, the magnitude of climate oscillations in the Holocene was different from the Late Würmian and decreased from the Early Holocene to the present.

Table 1 Information on the glacier oscillation of the Palü, Löbben and Göschener Cold Phases I/II

PALÜ OSCILLATION	**LÖBBEN OSCILLATION**	**GÖSCHENER COLD PHASES I/II**
± 9500 yr B.P. (ZOLLER et al., 1977)	± 3350-3150 yr B.P. (MAYR, 1964, 1968)	Phase I: ± 2830-2270 yr B.P. Phase II: ±1600-1200 yr B.P. (ZOLLER, SCHINDLER & RÖTHLISBERGER, 1966; ZOLLER, 1977)
Palynological evidence in the	*Palynological evidence in the*	
Swiss Central Alps:	**Swiss Central Alps:**	**Phase I:** = Alaphah-Mountain advance in Alaska (PORTER, 1964)
Hinterrhein Valley (BURGA, 1979, KISSLING, 1979) Upper Engadine (KLEIBER, 1974; HEITZ, 1982) Bernina Pass (BURGA, 1987) Puschlav Valley (ZOLLER et al., 1977; BURGA, 1987) Blenio Valley (MÜLLER, 1972)	Hinterrhein Valley (KISSLING, 1979) Oberhalbstein Valley (HEITZ, 1975) Prättigau Valley (WEGMÜLLER, 1976) Upper Engadine (BROMBACHER, 1981; PUNCHAKUNNEL, 1983; HEITZ, 1982) Puschlav Valley (BURGA, 1987) Blenio Valley (MÜLLER, 1972) Urseren Valley (KÜTTEL, 1982)	Palynological evidence: vegetational development of *Picea/Pinus cembra/Larix* forest ⇓ *Alnetum viridis* ⇓ *Epilobietum fleischeri*
Southern Alps:	**Bernese Oberland:**	**Phase II:**
Ganna-Varese (SCHNEIDER, 1985 = Ganna Oscillation)	Simmen Valley (WELTEN, 1952) Susten Pass (KING, 1974)	Palynological evidence: phases of forest recessions, high NAP values
Bernese Oberland:	**Valaisian Alps:**	
Faulenseemoos (EICHER, 1979)	Simplon Pass (LANG & TOBOLSKI, 1985)	
Valaisian Alps:		
Saas Fee (BURGA, 1982)		

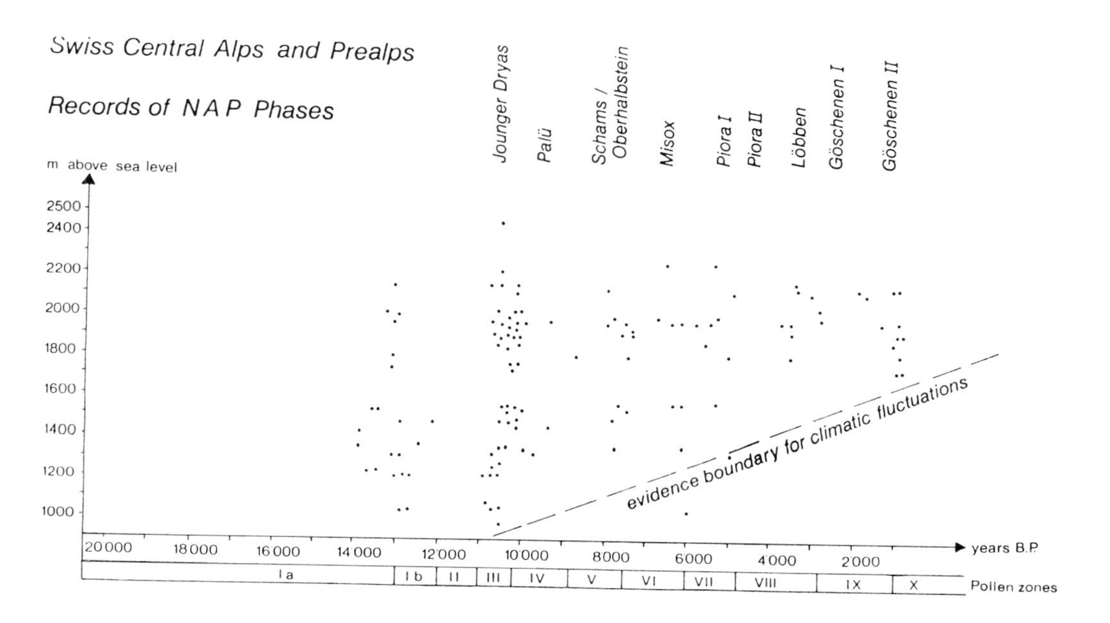

Fig. 3 Records of NAP phases: Swiss Central Alps and Prealps

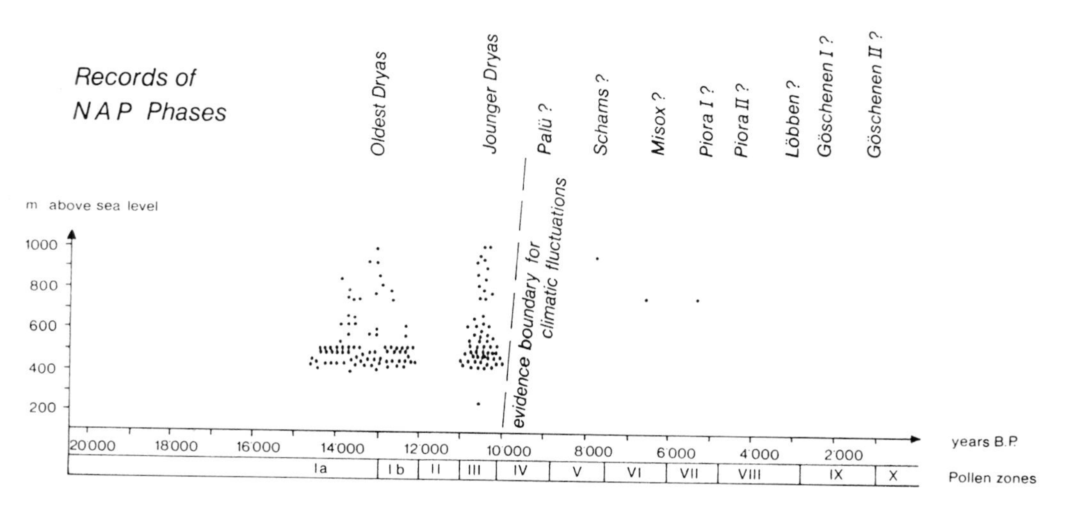

Fig. 4 Records of NAP phases: Jura Mountains and Swiss Plateau

References

AARIO, L. (1944): Über die pollenanalytischen Methoden zur Untersuchung von Waldgrenzen. Geol. Fören. Stockh. Förh. 66/3, 337-354

AMMANN, B. (1982): Säkulare Seespiegelschwankungen: wo, wie, wann, warum? Mitt. Naturf. Ges. Bern, N.F. 39, 97-106

BIGARELLA, J. J. & FERREIRA, A. M. M. (1985): Amazonian geology and the Pleistocene and Cenozoic environments and palaeoclimates. In: Prance, G. T. & Lovejoy, T. E. (eds.): Amazonia. Pergamon Press, Oxfort, 49-71

BIRKS, H. H. (1991): Holocene plant macrofossils from Vestspitsbergen, Norway. The Holocene 1

BOREL, J. L.; JORDA, M. & MONJUVENT, G. (1984): Variations climatiques, morphogénèse et évolution de la végétation postwurmiennes dans les Alpes françaises. Les Alpes, 43-53

BROMBACHER, C. (1981): Zur Vegetations- und Klimageschichte des Oberengadin. Pollenanalytische Untersuchungen am Moor Chalavus bei St. Moritz. Diplomarb. Botanisches Institut Univ. Basel

BROWN, I. M. (1990): Quaternary glaciations of New Guinea. Quat. Sci. Rev. 9, 273-280

BRUN, A. (1979): Recherches palynologiques sur les sédiments du Golfe de Gabès: résultats préliminaires. In: Géologie méditerranéenne 6/1, 247-264

BUDYKO, M. I. (1982): The Earth's climate: past and future. International Geophysics Ser. 29, Academic Press, London, 307 p.

BURGA, C. A. (1979): Postglaziale Klimaschwankungen in Pollendiagrammen der Schweiz. Viertelj.schr. N.G. Zürich 124/3, 265-283

BURGA, C. A. (1982): Zur Gletscher- und Klimageschichte des Saastales. Phys. Geogr. 9, 1-250

BURGA, C. A. (1984): Aktuelle Vegetation und Pollengehalt von Oberflächenproben der obermontanen bis subalpinen Stufe am Bernhardin-Pass (Graubünden/Schweiz). Jahresber. N.G. Graubünden 101, 53-99

BURGA, C. A. (1987): Gletscher- und Vegetationsgeschichte der Südrätischen Alpen seit der Späteiszeit. Denkschr. S.N.G. 101, Birkhäuser, Basel

BURGA, C. A. (1988): Swiss vegetation history during the last 18,000 years. New Phytol. 110, 581-602

BURGA, C. A. (1990): Vegetationsgeschichte und Paläoklimatologie. Viertelj.schr. N.G. Zürich 135/1, 17-30

BURGA, C. A. (1991): Vegetation history and palaeoclimatology of the Middle Holocene: pollen analysis of alpine peat-bog sediments, covered formerly by the Rutor Glacier, 2510 m (Aosta Valley, Italy). Global Ecol. Biogeogr. Letters 1, 143-150

CLAPPERTON, C. M. (1990): Quaternary glaciations in the Southern Hemisphere: an overview. Quat. Sci. Rev. 9, 299-304

DAMBLON, F. (1979): Les relations entre la végétation actuelle et les spectres polliniques sur le plateau des Hautes Fagnes (Ardennes, Belgique). Lejeunia 95, 65

EICHER, U. (1979): Die O18/O16- und C13/C12-Isotopenverhältnisse in spätglazialen Süßwasserkarbonaten und ihr Zusammenhang mit den Ergebnissen der Pollenanalyse. Diss. Univ. Bern

FANIRAN, A. & JEJE, L. K. (1983): Humid tropical geomorphology. Longman, London, 414 p.

FRANKENBERG, P. (1986): Zeitlicher Vegetationswandel und Vegetationsrekonstruktion des "neolithischen Klimaoptimums" in der Jeffara Südosttunesiens. Akad. der Wiss. Mainz, 83 p.

FRENZEL, B. (1967): Die Klimaschwankungen des Eiszeitalters. Vieweg, Braunschweig, 291 p.

FRENZEL, B. (1968): Grundzüge der pleistozänen Vegetationsgeschichte Nord-Eurasiens. Erdwiss. Forsch. 1, Steiner, Wiesbaden, 326 p.

FRENZEL, B. (1977): Postglaziale Klimaschwankungen im südwestlichen Mitteleuropa. Erdwiss. Forsch. 13, 297-322

GAILLARD, M. J. (1985): Postglacial palaeoclimatic changes in Scandinavia and Central Europe. A tentative correlation based on studies of lake level fluctuations. Ecologia Med. 11, 159-175

GRAF, K. J. (1981): Palynological investigations of Postglacial peat bogs near the boundary of Bolivia and Peru. Journal of Biogeography 8, 353-368

GRIBBIN, J. & LAMB, H. H. (1978): Climatic change in historical times. In: Gribbin, J. (ed.): Climatic change. Cambridge Univ. Press, 280 p.

GRICHUK, V. P. (1964): Comparative study of the interglacial and interstadial flora of the Russian Plain. Rep. VI. Int. Congr. Quat. 2, Warsaw 1961, 395-406

GRICHUK, V. P. (1984): Late Pleistocene vegetation history. In: Velichko, A. A.; Wright, H. E. & Barnowsky, C. W.: Late Quaternary environments of the Soviet Union. Longman, London, 155-178

GRICHUK, V. P. (1989): Floren- und Vegetationsgeschichte Rußlands im Pleistozän (in Russian). Nauka, Moskau, 184 p.

GROVE, J. H. (1979): The glacial history of the Holocene. Progr. Phys. Geogr. 3/1, 1-54

HAGER, P. K. (1916): Verbreitung der wildwachsenden Holzarten im Vorderrheintal (Kt. Graubünden). Dep. des Innern, Bern, 331 p.

HEIM, J. (1971): Etude statistique sur la valabilité des spectres polliniques provenant d'échantillons de mousses. Lejeunia 59, 34 p.

HEITZ, C. (1975): Vegetationsentwicklung und Waldgrenzschwankungen des Spät- und Postglazials im Oberhalbstein (Graubünden/Schweiz) mit besonderer Berücksichtigung der Fichteneinwanderung. Beitr. geobot. Landesauf. Schweiz 55, 1-63

HEITZ-WENIGER, A. K.; PUNCHAKUNNEL, P. & ZOLLER, H. (1982): Vegetations-, Klima- und Gletschergeschichte des Oberengadins. Phys. Geogr. 1, 157-170

HEUSSER, C. J. (1974): Vegetation and climate of the Southern Chilean District during and since the last interglaciation. Quat. Sci. Rev. 4, 290-315

HUNTLEY, B. & BIRKS, H. J. B. (1983): An atlas of past and present pollen maps for Europe: 0-13,000 years ago. Cambridge Univ. Press, Cambridge

IMBRIE, J. & PALMER IMBRIE, K. (1986): Ice Ages. Harvard Univ. Press, 224 p.

Isla, F. I. (1989): Holocene sea-level fluctuation in the Southern Hemisphere. Quat. Sci. Rev. 8, 359-368

Iversen, J. (1944): *Viscum*, *Hedera* and *Ilex* as climate indicators. Geol. Fören. Stockh. Förh. 66, 463-483

Jochimsen, M. (1972): Pollenniederschlag und rezente Vegetation in Gletschervorfeldern der Alpen. Ber. Deutsch. Bot. Ges. 85/1-4, 13-27

Jochimsen, M. (1986): Zum Problem des Pollenfluges in den Hochalpen. Diss. Bot. 90, 249 p.

Karner, A.; Kral, F. & Mayer, H. (1973): Pollenanalytische Untersuchungen zur Einwanderungsgeschichte der Tanne des Vintschgaues. In: Das inneralpine Vorkommen der Tanne im Vintschgau. Cbl. f. d. ges. Forstwesen 90, 129-163

Kellogg, W. W. (1978): Global influence of mankind on the climate. In: Gribbin, J. (ed.): Climatic change. Cambridge Univ. Press, 280 p.

King, L. (1974): Studien zur postglazialen Gletscher- und Vegetaionsgeschichte des Sustenpaßgebietes. Basler Beiträge zur Geographie 18, 1-125

Kissling, H. R. (1979): Das Pollendiagramm: Glaspaß. Diplomarb. Botanisches Institut Univ. Basel

Kleiber, H. (1974): Pollenanalytische Untersuchungen zum Eisrückzug und zur Vegetationsgeschichte im Oberengadin. Bot. Jahrb. Systematik 94, 1-53

Küttel, M. (1982): Züge der jungpleistozänen Vegetations- und Landschaftsentwicklung der Zentralschweiz. Phys. Geogr. 5, 33 p.

Kuhry, P. (1988): Palaeobotanical-palaeoecological studies of tropical high Andean peat-bog sections (Cord. Oriental, Colombia). Diss. Bot. 116, 241 p.

Kukla, G. J. (1978): Recent changes in snow and Ice. In: Gribbin, J. (ed.): Climatic change. Cambridge Univ. Press, 114-129

Lang, G. & Tobolski, K. (1985): Late Glacial and Holocene environment of a lake near the timberline in the Central Swiss Alps. Diss. Bot. 87, 209-228

Lézine, A. M. & Casanova, J. (1989): Pollen and hydrological evidence for the interpretation of past climates in tropical West Africa during the Holocene. Quat. Sci. Rev. 8/1, 45-55

Markgraf, V. (1980): Pollen dispersal in a mountain area. Grana Palyn. 19, 127-146

Markgraf, V. (1980a): Palaeoclimatic changes during the last 15,000 years in subantarctic and temperate regions of Argentina. Mem. Mus. Hist. Nat. 27, Paris

Markgraf, V. (1989): Palaeoclimates in Central and South America since 18,000 B.P. based on pollen and lake-level records. Quat. Sci. Rev. 8, 1-12

Mayr, F. (1964): Untersuchungen über Ausmaß und Folgen der Klima- und Gletscherschwankungen seit dem Beginn der postglazialen Wärmezeit. Z. Geomorph. N.F. 8/3, 257-285

Mayr, F. & Heuberger, H. (1968): Type areas of Late Glacial and Postglacial deposits in Tyrol, Eastern Alps. Proc. 7th INQUA Congr. Colorado Studies 7, 143-165

Müller, H. J. (1972): Pollenanalytische Untersuchungen zum Eisrückzug und zur Vegetationsgeschichte im Vorderrhein- und Lukmaniergebiet. Flora 161, 333-382

NEFTEL, A.; OESCHGER, H.; STAFFELBACH, T. & STAUFFER, B. (1988): CO_2-record in the Byrd ice core 50,000 - 5000 yr B.P. Nature 331, 6157, 609-611

PORTER, S. C. (1964): Late Pleistocene glacial chronology of North-Central Brooks Range, Alaska. Amer. Sci. 262

PORTER, S. C. & OROMBELLI, G. (1985): Glacier contraction during the Middle Holocene in the Western Italian Alps: evidence and implications. Geology 13, 296-298

PUNCHAKUNNEL, P. (1983): Pollenanalytische Untersuchungen zum Eisrückzug und zur Vegetationsgeschichte im Oberengadin II. Botanisches Institut Univ. Basel, B 66, 1-105

RÖTHLISBERGER, F. (1986): 10 000 Jahre Gletschergeschichte der Erde. Sauerländer, Aarau, 416 p.

ROGNON, P. (1983): Les crises climatiques de courte durée (Quelques années à quelques siècles) et leur enregistrement dans la sédimentation continentale. In: Ghazi, A. (ed.): Palaeoclimatic Research and Models. Reidel, Dordrecht, 114-123

RUDLOFF, H. VON (1980): Die Klimaentwicklung in den letzten Jahrhunderten im mitteleuropäischen Raume (mit einem Rückblick auf die postglaziale Periode). In: Oeschger, H.; Messerli, B. & Silvar, M. (eds.): Das Klima. Springer, Berlin, 125-148

SCHNEIDER, R. (1984): Vergleich des Pollengehaltes von Oberflächenproben mit der rezenten Vegetation im Aspromonte (Kalabrien, Italien). Diss. Bot. 72, 275-318

SCHNEIDER, R. (1984): Palynological research in the Southern and Southeastern Alps between Torino and Trieste. A review of investigations concerning the last 15,000 years. Diss. Bot. 87, 83-103

SCHNEIDER, R. & TOBOLSKI, K. (1985): Lago di Ganna: Late Glacial and Holocene environments of a lake in the Southern Alps. Diss. Bot. 87, 229-271

SELTZER, G. O. (1990): Recent glacial history and palaeoclimate of the Peruvian-Bolivian Andes. Quat. Sci. Rev. 9, 137-152

SUGGATE, R. P. (1990): Late Pliocene and Quaternary glaciations of New Zealand. Quat. Sci. Rev. 9, 175-197

TAUBER, H. (1977): Investigations of aerial pollen transport in a forested area. Dansk Bot. Ark. 32, 121 p.

TRIAT-LAVAL, H. (1978): Contribution pollenanalytique à l'histoire tardi- et postglaciaire de la végétation de la basse vallée du Rhône. Thèse Marseille III, 343 p.

WEGMÜLLER, S. (1977): Vegetationsgeschichteliche Untersuchungen in den Thuralpen und im Faningebiet (Kanton Appenzell, St. Gallen, Graubünden/Schweiz). Bot. Jahrb. Systematik 97 (2), 226-307

WEGMÜLLER, S. (1977): Pollenanalytische Untersuchungen zur spät- und postglazialen Vegetationsgeschichte der französischen Alpen (Dauphiné). Haupt, Bern

WELTEN, M. (1950): Beobachtungen über den rezenten Pollenniederschlag in alpiner Vegetation. Ber. Geobot. Inst. Rübel, 48-57

WELTEN, M. (1952): Über die spät- und postglaziale Vegetaionsgeschichte des Simmentals. Veröff. des Geobotanischen Instituts Rübel Zürich 26, 1-135

ZOLLER, H. (1977): Alter und Ausmaß postglazialer Klimaschwankungen in den Schweizer Alpen. Erdwiss. Forsch. 13, 271-281

ZOLLER, H.; SCHINDLER, C. & RÖTHLISBERGER, H. (1966): Postglaziale Gletscherstände und Klimaschwankungen im Gotthardmassiv und Vorderrheingebiet. Verh. Naturf. Ges. Basel 77, 97-164

Address of the author:

PD Dr. C. A. Burga, Geographisches Institut der Universität Zürich-Irchel, Winterthurerstraße 190; CH-8057 Zürich, Switzerland

Analysis of fossil stomata of conifers as indicators of the alpine tree line fluctuations during the Holocene

Brigitta Ammann & Lucia Wick

Summary

In order to trace tree line fluctuations during the Holocene various palaeobotanical-stratigraphical methods have been used: pollen-, macrofossil and wood analysis. Very few macro- and woodfossil studies exist in the Alps. In pollen diagrams the distinction between long-distance-transported and local pollen is crucial. WELTEN (1982) and his students identified conifers stomata in their pollen counts, thus a first approximation to needles as a sign of local presence. In the Valais, Central Alps, the two tree species forming the modern timberline, *Larix decidua* and *Pinus cembra* reforested the area at 2000-2100 m already in the Early Preboreal (shortly after 10,000 yr. B.P.). At most localities the pioneer *Larix* was somewhat earlier established than *Pinus cembra*. Around 6000 yr B.P. the tree line was at some sites as high as 2300 m a.s.l. From 5000 yr B.P. onwards tree lines were lowered; at several sites indicators of pasture or wood pasture are recorded. Further analysis of both surface samples and stratigraphic cores are needed to sharpen the tool of stomata analysis.

Zusammenfassung

Um die Lage fossiler Wald- und Baumgrenzen zu bestimmen, wurden als biostratigraphische Methoden die Analyse von Pollen, pflanzlichen Makroresten und Holz verwendet. Leider gibt es aber aus den Alpen bloß eine kleine Zahl von Makrorest- und Holzuntersuchungen aus der Nähe von Waldgrenzen. Für die Interpretation von Pollendiagrammen ist die Unterscheidung von Lokal-, Regional- und Fernflug von großer Bedeutung, da letzterer oberhalb der Waldgrenze überwiegen kann. WELTEN (1982) und seine Schüler pflegten regelmäßig Spaltöffnungen der Nadelhölzer in den Pollenpräparaten mitzuzählen, was für die Unterscheidung lokaler An- oder Abwesenheit von Nutzen ist. Für das Wallis konnten sie zeigen, daß die beiden Baumarten, die heute oft die Waldgrenze bilden, Lärche und Arve, bereits kurz nach 10 000 J.v.h. Höhenlagen von 2000-2100 m wieder bewaldeten. Meist siedelte sich die Pionierart Lärche etwas vor der Arve an. Um 6000 J.v.h. lag die Baumgrenze an mehreren Stellen auf 2300 m ü.M. Ab 5000 J.v.h. senkte sich die Lage der Baumgrenze. Wie weit die Ursachen dieser Erscheinung beim Klima oder beim menschlichen Einfluß liegen, muß lokal und in Zusammenarbeit mit der Archäologie bewertet werden. Mancherorts weisen Pollen, Holzkohle und gesteigerte Bodenerosion auf prähistorische Beweidung solcher waldgrenznahen Standorte hin. Die Untersuchung von Oberflächenproben im Hinblick auf Pollen, Makroreste und Stomata entlang ausgewählter Transekte durch die Wald- und Baumgrenzregionen sind notwendig, um die Methode der Stomata-Analyse zu verbessern.

1. Introduction

In alpine palaeoecology an important long-term goal is to understand the causes of Holocene timberline fluctuations - are they climatic or anthropogenic or both? If we could exclude human impact at specific sites or even in areas, we could then proceed in two directions:

(1) use the fluctuations for a reconstruction of past climates. This would be possible at least on a coarse scale if we bear in mind the inertia of tree populations as well as such non-climatic factors as slope processes, pedogenesis and topography;
(2) assess the reaction of the subalpine forests to climatic change, provided that we have independent evidence for climatic change. This will require multidisciplinary studies with high resolution in time and space.

Several steps towards this long-term goal involve actuo-ecology, e.g. ecophysiology and bioclimatology of the various species forming the altitudinal belt called the modern timberline. In addition we must remember the statement by TROLL (1973): "It is absolutely clear that upper timberlines in different parts of the world cannot be climatically equivalent, not even in a relatively small mountain system such as the Alps or Tatra mountains."

At the outset a distinction should be made between "tree line" and "timberline". The latter usually refers to the limit of closed forest, whereas the former describes the limit of trees (including Krummholz). Although the long-term goal of palaeoenvironmental tree line studies is to reconstruct the Holocene fluctuations, the more modest and more immediate objective is the seemingly simple question of what palaeoecological methods can be used to identify the location of past tree lines. Possible causes for Holocene tree line fluctuations can then be evaluated on the basis of results from parallel studies of charcoal influx, oxygen-isotope ratios in lacustrine carbonates, dendroclimatology, and archaeological surveys.

2. Palaeobotanical methods for identifying past tree lines

Several palaeobotanical-stratigraphical methods have been applied to peat and lake-sediments in order to evaluate whether a specific site was above or below tree line. The materials can be grouped as mega-, macro- or micro-fossils. Their advantages and disadvantages are summarized in Table 1.

Table 1 Advantages and disadvantages of palaeobotanical data for the reconstruction of timberlines

	production + dispersal	spatial resolution	taxonomic resolution	density of possible sites	density of existing studies	preservation
Megafossils: wood	low	high	to genus	medium	low	variable
Macrofossils: fruits, seeds, leaves, needles	low	high	often to species	medium	low	good
Microfossils: pollen, spores	high	low	to genus or family	good	good	very good
Stomata	low	high	to genus	good	low	good

The most widespread tool for tree line research is pollen analysis. In some polar regions the zone between tree line and timberline is very broad, so that bounderies are difficult to determine. The tree line in Northern Finland was first studied by simple pollen analysis of surface samples by AARIO (1940), but recent pollen-trap studies of polleninflux (HICKS, this volume) may be applied to stratigraphic sequences to determine the location of past tree lines (HYVÄRINEN, this volume).

In mountainous regions the alpine tree line may be much narrower locally than the polar tree line, but the complications introduced by topography, aspect, slope processes, soils, etc. commonly make it more irregular, and the distance transport of tree pollen from lower elevations blurs the relation of pollen percentages to the local tree line (see LÜDI & VARESCHI, 1936; MAHER, 1963; MARKGRAF, 1980).

In contrast to pollen analysis a more precise method for tree line reconstruction is the analysis of macrofossils (LANG, in press). Unfortunately a very small number of studies of sites near the timberline in the Alps is published. Efforts in this direction are urgently needed. This shortcoming may be alleviated by the use of stomata of conifers identified to genus in classical pollen analysis (MARKGRAF, 1969; WELTEN, 1982; WEGMÜLLER & LOTTER, 1990; WICK, 1989 in FEDELE et al., 1989, all based on the identification key by TRAUTMANN, 1953). In studies of reforestation during the Late Glacial in the lowlands stomata of *Juniperus* and *Pinus* are often recorded and interpreted as a sign of local presence rather than long distance transport (e.g. WELTEN, 1982; GAILLARD, 1984; AMMANN, 1989).

The value of stratigraphical stomata analysis will be improved when surface samples in transects across the tree line are investigated for pollen, macrofossils, and stomata.

3. Examples of stomata records in pollen studies in the Central Alps

WELTEN (1982) presented a north-south transect through the Bernese and Valais Alps based on over 30 pollen diagrams. Seven of them are selected here from both slopes of the longitudinal valley of the Valais (Fig. 1). (In addition the site Böhnigsee at 2095 m a.s.l. was described by MARKGRAF (1969), but is not included here because of problems in dating and correlation). Table 2 summarizes the vegetational belts of the seven sites.

The subalpine forests in this continental situation of the inner-alpine valley is not formed by *Picea excelsa* as on the northern slope of the Alps but by *Pinus cembra* (stone pine) and *Larix decidua* (European larch). The altitude of the potential climatic tree line is considered to be around 2000-2300 m a.s.l. The lakes and bogs analysed by WELTEN (1982) range from altitudes of 1510 m to 2290 m a.s.l. For his original diagrams WELTEN (1982) used various pollen sums by excluding taxa he considered as dominated by long distance transport; the selection of taxa included in the pollen sum was therefore different at different sites, according to their altitude. For the present paper a uniform pollen sum was calculated that includes all trees, shrubs, and herbs but excludes Cyperaceae, water plants, and ferns. The stomata were recorded by WELTEN (1982) in numbers per pollen count. We express these abundancies in Figs. 2-6 as percentages of the pollen sum.

The data since the end of the Younger Dryas is included here except for the sites Eggen-Blatten and Aletschwald which only begin in the Preboreal, and Grächen which starts with the Older Atlantic. Zeneggen-Hellelen and Simplon-Hopschensee reach back to the Oldest Dryas and the Allerød, respectively, but these intervals are not included here.

Pinus (Fig. 2)

Unfortunately the stomata of the three species involved cannot be differentiated. In pollen *Pinus cembra* can be distinguished, and *P. silvestris* and *P. mugo* are grouped together in *Pinus* non-*cembra*.

At all elevations recorded the *Pinus* non-*cembra* pollen is abundant already in the Younger Dryas. *Pinus* stomata occur this early only at the lowest site; high *Pinus cembra* pollen and the absence of stomata favour the interpretation of strong transport of *Pinus* pollen from below. From the Preboreal onwards *Pinus* stomata occur at all elevations, except for the highest Mont Carré, where stomata records start during the Boreal. They are particularly common at Simplon-Hopschensee, where the location at the pass may be favourable.

At the transition from Boreal to Older Atlantic the *Pinus cembra* pollen percentages increase at Aletschwald, Hopschensee, and Mont Carré. Simultaneously the occurrence of

Pinus stomata becomes very regular: *Pinus cembra* pollen shows several younger peaks at intermediate altitudes (1710 m, 1645 m) in the Atlantic and the Subbboreal but is absent from the lowest site (1510 m, today in the belt of subalpine spruce forests with some stone pines).

The decrease in *Pinus cembra* pollen and stomata in the Subboreal and Subatlantic coincides with signs of human influence in the lowland (pollen of Cerealia and weeds). WELTEN (1950, 1982) discussed the deforestation and/or the transformation to wood pastures.

In the WELTEN dataset the minimum pine pollen percentage with stomata is 15%. The maximum pine pollen percentage without stomata is 89%.

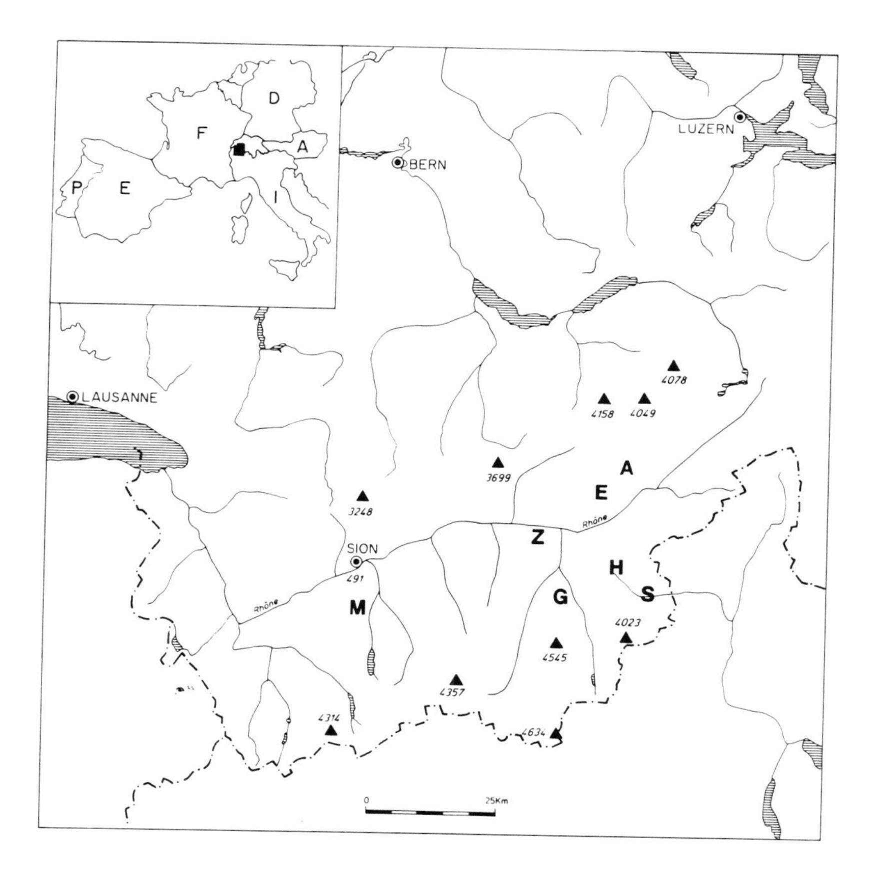

Fig. 1 Map of the Valais with seven sites studied by WELTEN (1982): A=Aletschwald, E=Eggen (near Blatten), G=Grächen, H=Hopschensee (on Simplon); M=Mont Carré, S=Alter Spittel, Z=Zeneggen (Hellelen). Mountain peaks with altitudes in m

Table 2 Sites of pollen diagrams in the Valais Alps (WELTEN, 1982)

Site	Altitude in m a.s.l.	Vegetation potential natural	today
Mont Carré	2290	alpine meadows	alpine meadows (pastures)
Aletschwald	2017	*Larix & Pinus cembra* forest	*Larix & Pinus cembra* forest
Hopschensee	2017	*Larix & Pinus cembra* forest	alpine meadows (pastures)
Alter Spittel	1885	*Larix-Picea-Pinus cembra* forest	alpine meadows (pastures)
Grächensee	1710	*Larix-Pinus cembra-Pinus sylvestris-Picea* forest	decades ago: cereals today: meadows (pastures)
Eggen-Blatten	1645	*Picea-Larix-Pinus cembra* forest	alpine meadows (pastures)
Zeneggen-Hellelen	1510	transition from subalpine (*Picea-Larix-Pinus cembra*) to upper montane forest (*Pinus sylvestris-Picea-Abies*)	today upper limit of cereals; (pastures) and hay-meadows

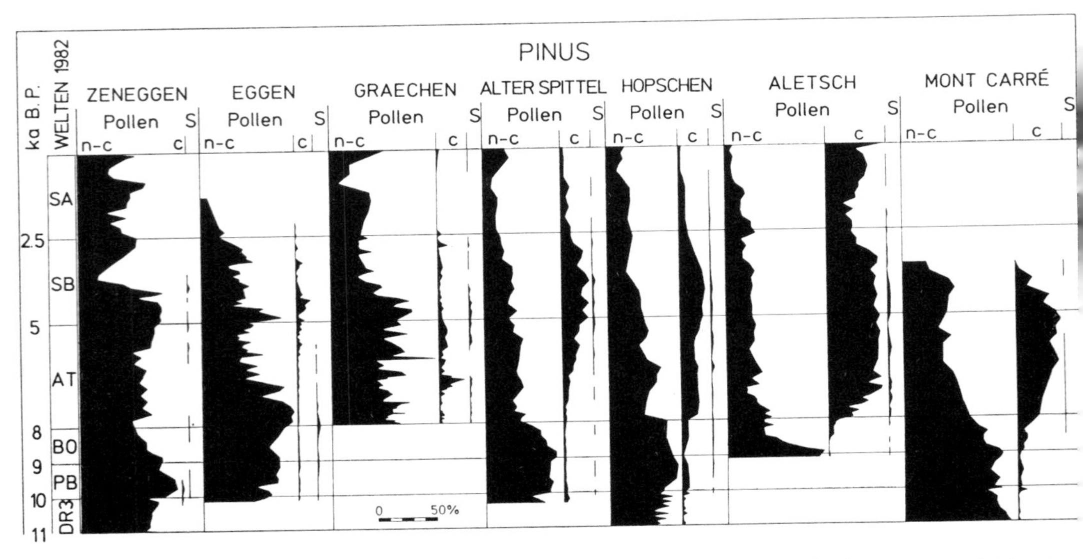

Fig. 2 Pinus: Pollen percentages and stomates (as % of pollen sum) from the lowest to the highest site of WELTEN (1982). n-c=*Pinus* non-*cembra*, c=*Pinus cembra*. S=stomates

Juniperus (Fig. 3)

Juniperus which plays a major role in reforestation during the Bølling at low elevation (e.g. BERTSCH, 1961) including Zeneggen-Hellelen, is also important in the Early Holocene over a wide range of elevations. Similarly DE BEAULIEU (1977) showed such a double role of juniper across Europe.

In the WELTEN (1982) dataset the minimum pollen percentage for *Juniperus* in the Early Holocene with stomata is 0%. The maximum pollen percentage without stomata is 17%.

Larix (Fig. 4)

Larix is a very poor pollen producer but a great producer of needles and stomata. Often the first occurrence of pollen and stomata is recorded in the same sample and only rarely does the pollen curve show a long "tail", i.e. single grains before the continuous curve. At two sites (Hopschensee and Böhnigsee) stomata were found before pollen grains.

From the very beginning of the Preboreal *Larix* is present (pollen and stomata) at all sites considered except for the highest one, Mont Carré, where it immigrated during the Preboreal. For the Early Holocene the abilities of *Larix* as a pioneer tree are illustrated by the diagrams for Alter Spittel, Hopschensee, Aletschwald, and Mont Carré; at all these sites *Larix* stomata were found before *Pinus* stomata.

For the later Holocene *Larix* keeps more or less similar values at the sites Grächen, Aletschwald, and Alter Spittel; in the first two diagrams a young regeneration of *Larix* is observed. At the sites Zeneggen-Hellelen, Eggen-Blatten, Hopschensee, and Mont Carré a reduction of *Larix* pollen and a lack of *Larix* stomata late in the Atlantic point to the disappearance of the tree, which is often interpreted as a result of human impact (synchroneity with indicators of pasture).

The close correlation between percentages of *Larix* pollen (which does not suffer from long distance transport like pine pollen) and *Larix* stomata is convincing evidence that conifer stomata are a reliable indicator for local presence of the source tree.

In the WELTEN dataset the minimum pollen percentage with stomata is 0%. The maximum pollen percentage without stomata is 23%.

Picea (Fig. 6)

Picea normally shows a long "tail" of single pollen grains or low values (i.e. absolute limit and empirical limit) before a distinct rise that can be taken as the rational limit. At many sites this step in the pollen curve is accompanied by the first occurrence of stomata

(Zeneggen-Hellelen, Grächen, Hopschensee). In two cases the first stomata are found somewhat later (Alter Spittel and Aletschwald). At Mont Carré and Eggen-Blatten no stomata of *Picea* are recorded. This is natural for Mont Carré because of its high elevation; but for Eggen-Blatten this absence of stomata is surprising.

Picea expansion was sometimes explained as favored by prehistoric human influence (WELTEN, 1952, 1982; MARKGRAF, 1970, 1972) as suggested by the synchroneity of the pollen increase with indicators for pasture. In the Subatlantic *Picea* pollen decreases and stomata become very rare (except for Alter Spittel, which is open to pollen rain from the southern slopes of the Alps).

In the WELTEN dataset the minimum pollen percentage of *Picea* with stomata is 3%. The maximum pollen percentage without stomata is 56%.

Abies (Fig. 5)

Abies is of restricted importance in the sub-continental climate of the Valais. It occurs today occasionally between the dry pine forests near the valley bottom and the lower subalpine forest with *Picea excelsa*. *Abies* never forms the tree line. In the WELTEN dataset *Abies* stomata occur only at three sites, Eggen, Grächen, and Aletschwald, mainly during

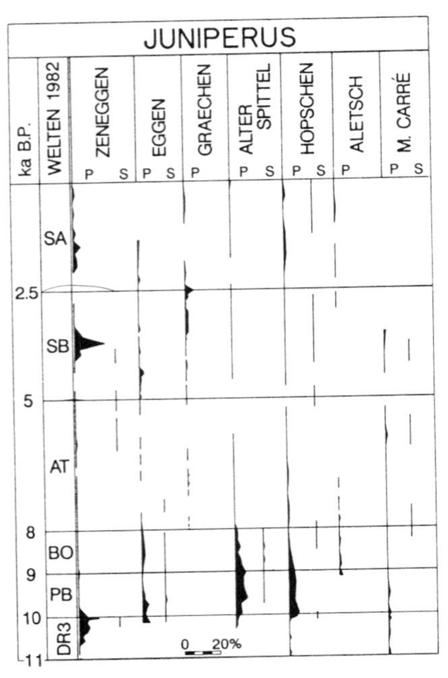

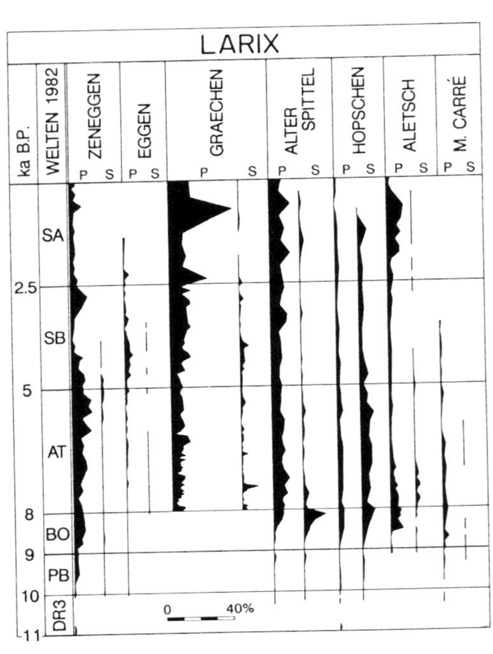

Fig. 3 / 4 *Juniperus* and *Larix*: pollen percentages and stomates (as % of pollen sum) from the lowest to the highest site of WELTEN (1982). P=pollen, S=stomates

the Atlantic, and at the lowest site Eggen it is found during the Subboreal. Eggen and Grächen are at altitudes within the modern occurrence of *Abies* but the finds at Aletschwald (2017 m a.s.l.) are rather surprising. At all sites pollen percentages are low and stomata are absent during the last 2500 years.

In the WELTEN dataset the minimum pollen percentage with stomata is 3%. The maximum pollen percentage without stomata is 33%.

4. Conclusions

Both WELTEN (1982) and MARKGRAF (1969) give a detailed discussion of the tree line fluctuations at their sites in the Valais on the basis of the occurrence of both pollen and stomata. The presence or absence of stomata helps to assess the importance of transported pollen from below the tree line.

From four localities above 2000 m a.s.l. (Hopschensee, Aletschwald, Mont Carré of WELTEN, 1982 and Böhnigsee of MARKGRAF, 1969) the reforestation up to at least 2000 m dates back to the beginning of the Preboreal, shortly after 10,000 yr B.P. Around 6000 yr B.P. tree lines seem to rise to 2100-2300 m: Mont Carré 2290 m had a stable *Pinus cembra* forest. From 5000 yr B.P. onwards tree lines were considerably lower - the question opens whether this resulted from human impact, from climate, or from both.

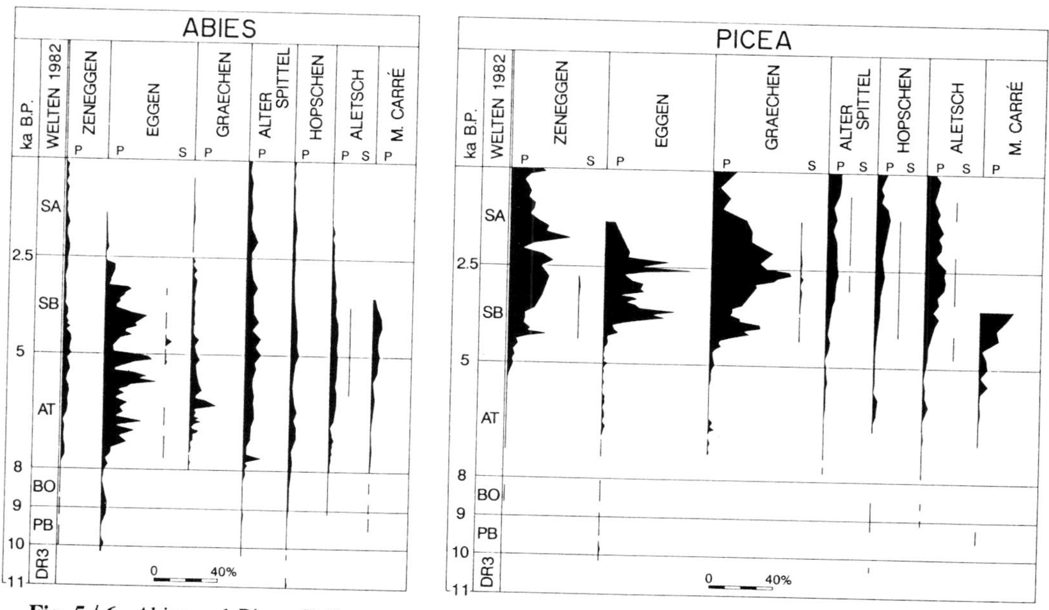

Fig. 5 / 6 *Abies* and *Picea*: Pollen percentages and stomates (as % of pollen sum) from the lowest to the highest site of WELTEN (1982). P=pollen, S=stomates

Combined analysis of charcoal, lithostratigraphy, and palaeomagnetism (on the same cores) together with an archaeological survey of the area may give evidence of human impact.

Such multidisciplinary study is currently undertaken in the Italian Alps near the pass of Splügen. WICK (in FEDELE et al., 1989) shows in her preliminary results from Lago Basso (2250 m a.s.l.) and Lago Grande (2303 m) that shortly before 5000 yr B.P. both pollen and stomata of the two timberline taxa *Larix decidua* and *Pinus cembra* begin to decrease, whereas the avalanche tolerant shrub *Alnus viridis* increases, implying reduction of tree cover. After several steps of decline the stomata of both trees disappear at a level that is characterized by the new crops of the Roman period (i.e. *Juglans regia* and *Castanea sativa* pollen) and by a sharp rise in NAP, especially Gramineae, *Urtica*, *Plantago*, Chenopodiaceae, Cichoriaceae, and Asteraceae. At the same level an abrupt increase of clay in the gyttja is observed. These are all indications that summer farming was intensified with the increasing human population around 2000 yr B.P., and that the tree line dropped locally below 2250 m.

In order to determine the elevation of past tree lines we need to include systematic identification and enumeration of stomata in our pollen analysis. To support the interpretation of stratigraphic profiles of pollen/stomata diagrams, we also need studies of surface samples in altitudinal transects from below timberline to above tree line with pollen, macrofossils, and stomata all quantified and related to phytosociological relevées to that transition.

Acknowledgements

We gratefully acknowledge the help of B. Brogli, J. Gratzfeld, M. Kummer, O. Schläfli and H. E. Wright Jr.

References

AARIO, L. (1940): Waldgrenzen und subrezente Pollenspektren in Petsamo, Lappland. Publ. Inst. Geogr. Univ. Helsing. 3, 120 p.

AMMANN, B. (1989): Late Quaternary palynology at Lobsigensee. Regional vegetation history and local lake development. Diss. Bot. 137, 157 p.

BEAULIEU, J.-L. DE (1977): Contribution pollenanalytique à l'histoire tardiglaciaire et holocène de la végétation des Alpes méridionales françaises. Thèse Marseille III, 358 p.

BERTSCH, A. (1961): Untersuchungen zur spätglazialen Vegetationsgeschichte Südwestdeutschlands. Flora 151, 243-280

FEDELE, F.; KVAMME, M.; MOTTURA, A. & WICK, L. (1989): Preistoria e palaeoambienti della Valchiavenna, Campagna di ricerche 1989: Pian dei Cavalli, Baldiscio e Spluga. Clavenna 28, 9-78

GAILLARD, M.-J. (1984): Etude palynologique de l'évolution tardi- et postglaciaire de la végétation du moyen-pays Romand (Suisse). Diss. Bot. 77, 322 p.
HICKS, S. (1992): Investigating natural and anthropogenic changes in the polar tree limit in Northern Fennoscandia. (This volume)
HYVÄRINEN, H. (1992): Holocene pine and birch limits near Kilpisjärvi, Western Finnish Lapland: pollen stratigraphical evidence. (this volume)
LANG, G. (in press): Holozäne Veränderungen der Waldgrenze in den Schweizer Alpen - Methodische Ansätze und gegenwärtiger Kenntnisstand. Dissertationes Botanicae
LÜDI, W. & VARESCHI, V. (1936): Die Verbreitung, das Blühen und der Pollenniederschlag der Heufieberpflanzen im Hochtale von Davos. Ber. Geobot. Forschungsinst. Rübel in Zürich für 1935, 47-112
MAHER, L. J. Jr. (1963): Pollen analyses of surface materials from the southern San Juan Mountains, Colorado. Geol. Soc. Am. Bulletin 74, 1485-1504
MARKGRAF, V. (1969): Moorkundliche und vegetationsgeschichtliche Untersuchungen an einem Moorsee an der Waldgrenze im Wallis. Bot. Jahrb. 89, 1-63
MARKGRAF, V. (1970): Palaeohistory of the spruce in Switzerland. Nature 228, 249-251
MARKGRAF, V. (1972): Die Ausbreitungsgeschichte der Fichte (*Picea abies* H. Karst) in der Schweiz. Ber. Dt. Bot. Ges. 85, 165-172
MARKGRAF, V. (1980): Pollen dispersal in a mountain area. Grana 19, 127-146
TRAUTMANN, W. (1953): Zur Unterscheidung fossiler Spaltöffnungen der mitteleuropäischen Coniferen. Flora 140, 523-533
TROLL, C. (1973): The upper timberlines in different climatic zones. Arct. Alp. Res. 5/2, A3-A18
WEGMÜLLER, S. & LOTTER, A. F. (1990): Palynostratigraphische Untersuchungen zur spät- und postglazialen Vegetationsgeschichte der nordwestlichen Kalkvoralpen. Bot. Helv. 100, 37-73
WELTEN, M. (1952): Über die spät- und postglaziale Vegetationsgeschichte des Simmentals. Veröff. Geobot. Inst. Rübel 26, 135 p.
WELTEN, M. (1982): Vegetationsgeschichtliche Untersuchungen in den westlichen Schweizer Alpen: Bern-Wallis. Denkschr. S.N.G. 95, 104 p., 37 diagrams

Addresses of the authors:

Prof. Dr. B. Ammann, Botanische Institute und Botanischer Garten, Universität Bern, Altenbergrain 21, CH-3013 Bern, Switzerland
Dr. L. Wick, Botanische Institute und Botanischer Garten, Universität Bern, Altenbergrain 21, CH-3013 Bern, Switzerland

Late Quaternary forest line oscillations in the West Carpathians

Eliška Rybníčková & Kamil Rybníček

Summary

The existing views on the reconstruction of the forest limit oscillations in the West Carpathians, namely in the High Tatra region, during the Late Quaternary are reported on in this article. Macroscopic and palynological records of *Pinus cembra-Larix europaea-Juniperus* and *Salix* assemblages are used as an indication of the forest limit vegetation and, consequently, for the determination of the upper forest limit line. This type of vegetation covered the forelands of the West Carpathians between 100-500 m a.s.l. during the Late Pleistocene (40,000-20,000 yr B.P.). In the Late Glacial (20,000-10,000 yr B.P.) the forest limit vegetation invaded the intramontane basins up to about 700 (800?) m a.s.l. *Pinus cembra* and the accompanying trees disappeared from the lowlands and basins at about 9500 yr B.P., and rose to above 1000 m in the Tatras during the Preboreal and Boreal. No evidence of heliophytic forest limit vegetation has been found in the sediments of the climatic optimum. It is probable that a dense spruce forest reached its edaphic limit between 1900-2000 m a.s.l. at that period. The reappearance of the *Pinus cembra* assemblage from scattered relic sites and the beginning of the decrease of the forest line to altitudes of about 1650 m a.s.l. are dated back to about 4000 yr B.P. The present mean forest limit (1450 m a.s.l.) is anthropogenically lowered by about 120-150 m.

Zusammenfassung

In diesem Artikel werden die jüngeren Ansichten über die Schwankungen der Waldgrenze während des Spätquartärs in den Westkarpaten, namentlich in der Hohen Tatra, dargestellt. Als Indikatoren für die Vegetation der Waldgrenze werden die in den Sedimenten gefundenen Großreste und Pollenkörner der *Pinus cembra-Larix europaea-Juniperus* und *Salix*-Gesellschaften herangezogen. Dieser Vegetationstyp bedeckte während des Spätpleistozäns (40,000-20,000 J.v.h.) die Vorgebirgsebene der Westkarpaten. Im Spätglazial (bis ca. 10,000 J.v.h.) ist dieser Vegetationstyp in die innerkarpatischen Becken (bis etwa 700-800 m Höhe) vorgedrungen. Aus den Tiefebenen und Becken ist diese Waldgrenzgesellschaft etwa um 9500 J.v.h. verschwunden, und erreichte während des Präboreals und des Boreals vereinzelt Höhen von bis zu 1000 m ü.N.N. Erneut erscheint die *Pinus cembra*-Gesellschaft ausgehend von den zerstreuten Reliktstandorten um etwa 4000 J.v.h. und die Waldgrenze beginnt auf rund 1600 m zu sinken. Die heutige, anthropogen um etwa 120-150 m herabgedrückte Waldgrenze liegt in der Hohen Tatra bei einer Meereshöhe von 1450 m.

1. Introduction

In the Carpathians, the reconstructions of the forest limit oscillations during the Late Quaternary are, were, and will for a long time be of a highly hypothetical character. Compared to the Alps, there are not as many possibilities and data by far to determine the forest and timberline changes. All the glaciers melted a long time ago, at least 10,000-9000 yr B.P., and the traces of their Holocene oscillations, represented in the Alps, e.g. with several younger moraines, cannot be used for our purpose in the Tatras, simply because they do not exist. Palaeobotanical evidence is still incomplete and scattered especially from the uppermost parts of the mountains; nevertheless, it is the only evidence which can be used at all. A survey of the existing opinions on forest line changes, and a summary of all existing palaeobotanical and palynological evidence which could be used are presented here.

2. Present opinions on forest line oscillations

The current opinions on forest line oscillations, based mostly on palynological data in the High Tatras, were recently collected by KRUPIŃSKI (1984). He used many papers by Polish authors (KOPEROWA, 1958, 1962; FABIJANOWSKI, 1962; HESS, 1968; SZAFER, 1952, 1966; RALSKA-JASIEWICZOWA, 1972, and RALSKA-JASIEWICZOWA & STARKEL, 1975) and at the same time presents his own opinion. According to the oscillation curve published in his paper, the forest line did not exceed 250 m in the Dryas 1, 750-800 m in the Bølling, 550 m in the Dryas 2, 1100 m in the Allerød, 750 m in the Dryas 3, 1100 m in the Preboreal, 1650 m in the Boreal, 1850-1950 m in the Atlantic, 1750 m in the Subboreal, and 1550 m, which is the height of the present mean forest line, in the Subatlantic. KRUPIŃSKI's scheme of forest line oscillations in the High Tatras corresponds very closely with conclusions presented by FIRBAS (cf. WALTER & STRAKA, 1970: 181) for the Central European mountains in general.

3. Methodological base

Our own model of forest line oscillations is based on those palaeobotanical papers which document the presence and frequency of pollen and macroscopic remains of *Pinus cembra* or some other trees usually accompanying it. *Pinus cembra* is at present the most significant tree for the forest limit vegetation belt (=upper forest and lower subalpine belts) in the High Tatras. It grows there in combination with *Pinus mugo, Larix europaea, Picea abies, Juniperus communis* ssp. *nana*, *Salix* spec. div., dwarf shrubs (*Empetrum hermaphroditum, Vaccinium myrtillus, Vaccinium vitis-idaea, Vaccinium uliginosum*), grasses (*Calamagrostis villosa, Deschampsia flexuosa*), herbaceous plants (*Mulgedium alpinum, Senecio fuchsii* agg., *Solidago virga-aurea, Doronicum austriacum, Veratrum lobelianum, Adenostyles alliariae*), many other mountain plants (*Luzula sylvatica, Homogyne alpina,*

Soldanella carpatica, etc.), several mosses, lichens, and some ferns. Therefore, we took the pollen assemblage of *Pinus cembra-Larix-Pinus* t. *sylvestris* (incl. *Pinus mugo*)-*Juniperus-Salix* as an indication of the forest limit vegetation, which had already existed in the past.

4. A survey of past *Pinus cembra* assemblage sites

Published pollen assemblages of the *Pinus cembra* type have been looked up. They come from different parts of the West Carpathians, especially from their central parts (the Tatra Mountains and their forelands). Most of them are ^{14}C dated; cp. Fig. 1 for their location. The oldest dated finds of the above mentioned pollen assemblage stem from the northern foreland of the West Carpathians. From the old river terraces at Brzeźnica (1) in Poland at about 100 m a.s.l. MAMAKOWA & STARKEL (1974) reported that the *Pinus cembra* pollen assemblage dates back to about 36,000 yr B.P. Corresponding data exist from Myslenice (2) at about 300 m a.s.l. and Dobra (3) at about 470 m a.s.l. dated by ŚRODOŃ (1968) back to about 38,000 and 32,000 yr B.P. respectively. *Pinus cembra* forest limit vegetation is not documented in altitudes above about 500 m a.s.l. (cf. ŚRODOŃ, 1968).

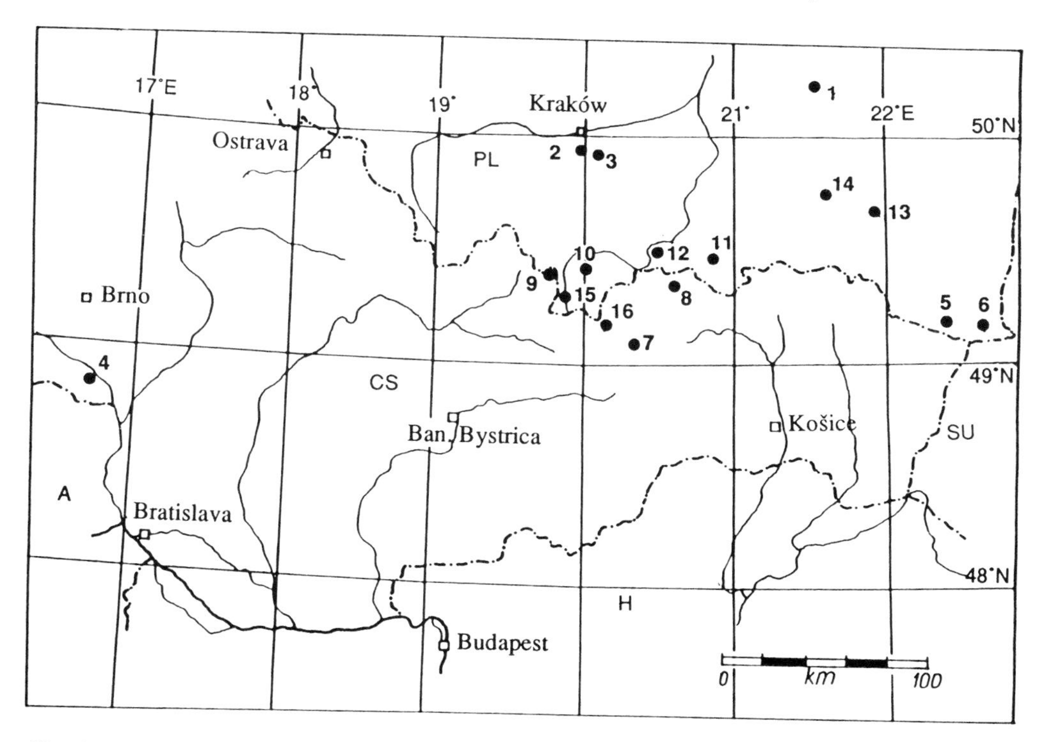

Fig. 1 The region of the West Carpathians and their foreland. Black dots: location of palaeoecological sites mentioned in the text and in Fig. 2 (for further explanations see chapter 4)

From the western margin of the Carpathians we found the *Pinus cembra* pollen assemblage in an old peat buried under 10 m of loess in South Moravia at Bulhary (4) at about 200 m a.s.l. It is dated to about 26,000 yr B.P. (HAVLIČEK & KOVANDA, 1985; RYBNÍČKOVÁ & RYBNÍČEK, 1989). Close by corresponding wood remains (charcoals) were found in palaeolithic excavations, dated to about the same age (KNEBLOVÁ, 1954).

RALSKA-JASIEWICZOWA (1980) published the *Pinus cembra* pollen assemblage from the Bieszczady Mountains. She found it at Smerek (5) and Tarnawa Wyżna (6) at an altitude of 600 and 670 m a.s.l. respectively. It is dated to about 15,000 yr B.P. at Smerek and at least 12,000 yr B.P. at Tarnawa Wyżna. A remarkable decrease of the pollen curves of *Pinus cembra, Larix,* and *Juniperus* can be observed during the Dryas 3 between about 10,800 and 10,300 yr B.P. A short increase then follows during the Preboreal and very rapid extinction of all those trees between about 9500 and 9000 yr B.P.

Similar pollen analytical results were obtained by JANKOVSKÁ (1988) from the Poprad Basin (7) in the southeastern foothills of the High Tatras at an altitude of about 690 m a.s.l. On another site at Sivárna (8) at about 600 m a.s.l. she even found *Pinus cembra* seeds and *Larix* cones ^{14}C dated to 11,340 yr B.P. (JANKOVSKÁ, 1984).

On the eastern side of the High Tatras we found pollen of *Pinus cembra, Larix*, and *Juniperus*, dated between 11,000 and 9500 yr B.P. in the Orava Basin (9). The low pollen occurrences and final extinction at 9500-9000 yr B.P. correspond to the situation observed in another intramontane basin (RYBNÍČKOVÁ & RYBNÍČEK, 1989) i.e. in the neighbouring Nowy Targ Basin in Poland. Pollen and macroscopic remains of *Pinus cembra, Pinus mugo*, and *Larix* are reported from Grel (10) by KOPEROWA (1958, 1962) from Bølling to Dryas 3, decline and extinction again happened in the Preboreal. The altitude of the Orava and Nowy Targ Basins is about 600 m a.s.l.

A few pollen grains of *Pinus cembra* and *Larix* were found by PAWLIKOWA (1965) in layers of the Dryas 3 and the Preboreal from altitudes of about 600 m in the Pieniny region (11). But ŚRODOŃ (1952) found macroscopic remains of both trees in Late Glacial samples nearby (12). This means that both species were more frequent there than could be supposed from the isolated determinations of pollen by the former author.

The *Pinus cembra* pollen assemblage also occurs in pollen diagrams from the Jaslo-Sanok Depression at the altitude of about 200 m a.s.l. (13, 14). Pollen of the present timberline trees occurred in great quantities (over 20% total sum) before 11,000 or 9500 yr B.P. as reported by KOPEROWA (1970) and HARMATA (1987) respectively. The extinction of the *Pinus cembra* assemblage took place there, according to the dates in both papers, between 9500 and 8500 yr B.P. A great difference in dating causes a certain hesitation in this case, the radiocarbon data by HARMATA seems to be a little too young.

5. A reconstruction of the forest limit oscillations

The result of our research is presented schematically in Fig. 2, which, at the same time, gives the new conception about the forest limit oscillations since about 40,000 yr B.P.

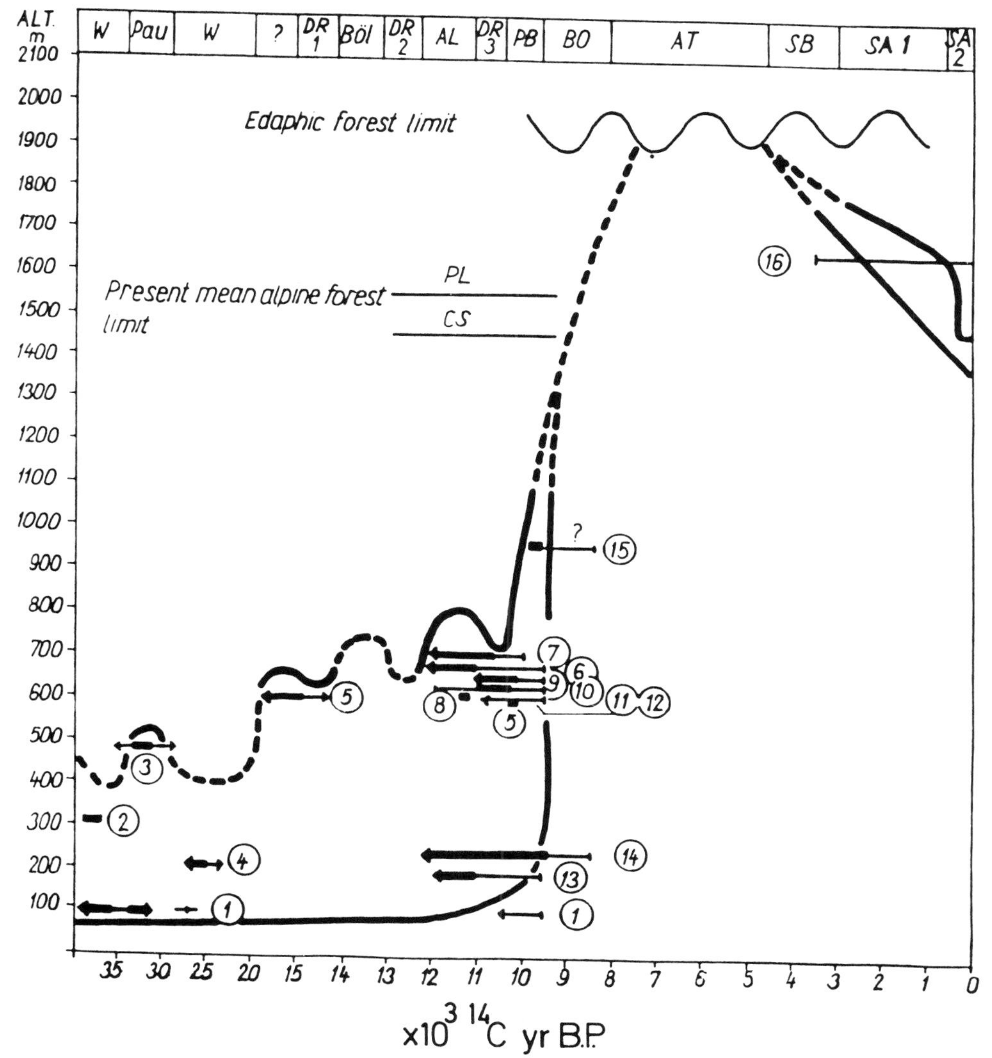

Fig. 2 Forest tree line oscillations in the West Carpathians (upper and lower limit of the *Pinus cembra* assemblage). Full line: documented level; dashed line: expected course of the upper limit. Numbers of localities correspond to numbers in Fig. 1 and in chapter 4.

All the data mentioned in chapter 4 inform us that the type of vegetation we know from the present forest limit belt covered the forelands of the Western Carpathians during the Upper Pleistocene. In the intramontane basins of the central part of the West Carpathians this type was common during the Late Glacial. We can observe a rising tendency of the forest limit line between at least 40,000 and 11,000 yr B.P., which was interrupted only by short temporary decreases during climatic deteriorations. A very rapid rise of the forest line could be reconstructed after the last cold Late Glacial period, the Dryas 3, i.e. after 10,000 yr B.P. The extinction of open coniferous stands with *Pinus cembra, Larix, Pinus mugo*, and *Juniperus* took place in the whole area of their previous occurrence in the lower parts of the Carpathians at about 9500 yr B.P. Instead, this type of vegetation is reported from the High Tatras at and above altitudes of about 1000 m a.s.l. (15) by KOPEROWA (1962) for the Preboreal; a scattered or isolated appearance can still be observed during the Boreal at these altitudes.

Since that time there has been no evidence of the occurrence of *Pinus cembra* in the West Carpathians, not even in the highest altitudes. The pollen diagram by HÜTTEMANN (HÜTTEMANN & BORTENSCHLAGER, 1987) from the Trojrohé pleso mire (16) in the High Tatras (1665 m a.s.l.) reaches back to about 6500 yr B.P. and does not show any pollen grains of *Pinus cembra* during the Atlantic and the lower part of the Subboreal periods. The pollen grains of *Larix* are very rare and isolated. The lack of *Pinus cembra* and the retreat of *Larix* during the Holocene climatic optimum can only be explained with the non-existence of the forest line vegetation belt in the High Tatras. The dense spruce forest probably reached the edaphic limit of its possible existence at about 1900-2000 m a.s.l. This is the very altitude of rock peaks, granite block screes, and sites with no continuous soil cover, forming the highest parts of the mountains up to 2655 m (the highest point is the Gerlach Mountain). The upper forest belt was certainly formed with spruce (*Picea abies*), very probably with *Corylus* shrubs and with some *Ulmus, Acer*, and *Sorbus* during the climatic optimum. We suppose that *Pinus cembra* and other heliophytic trees did not disappear from the area but survived the forest period in extreme sites (rocks and stabilized screes) as more or less isolated specimens or as an admixture in the uppermost forest stands. Outside the Tatra Mountains, however, *Pinus cembra* disappeared definitely in the Preboreal and has never spread again.

The reappearance of *Pinus cembra* pollen, followed by the rise of *Larix* and *Salix* pollen curves is dated in HÜTTEMANN's diagram (HÜTTEMANN & BORTENSCHLAGER, 1987) back to about 3500 yr B.P. We expect that the natural decrease of the forest and its upper limit started at about 4500 yr B.P. in the High Tatras and enabled a new spreading of *Pinus cembra, Larix, Juniperus, Pinus mugo* and other heliophytic subalpine plants since that time. Nevertheless, the representation of *Pinus cembra* and *Larix* has never reached their previous importance which was observed in the Carpathians during the Late Glacial and earlier.

On the Slovak side of the High Tatras the present forest limit with *Pinus cembra* and *Larix* oscillates around 1450 m (PLESNIK, 1971), depending on the habitat conditions. From the Polish part of the High Tatra Mountains the mean value is reported to be about 1550 m (KRUPIŃSKI, 1984). However, it is well documented that the natural forest limit line lies about 120-150 m higher in the whole Tatra system. It was artificially depressed due to summer grazing and tree cutting, namely the selection of *Pinus cembra* and *Larix* timbers, during the last 400-500 years.

6. Conclusions

To fill the existing gaps in our knowledge, we need to get more data especially from the highest parts of the High Tatras, from the raw humus layers, and the lake sediments situated close to the edaphic limit of the forest growth. Other pollen analyses from mires at and around the present forest limit would be of great help. The precise determination of pollen of *Pinus cembra, Larix, Juniperus*, and other characteristic plants is, together with reliable radiocarbon datings, the necessary premise for the studies of forest limit oscillations.

References

FABIJANOWSKI, J. (1962): Lasy tatrzańskie. In: Szafer, W. (ed.): Tatrzański park narodowy. Wydawnictwo popularno naukowe 10, 240-304

HAVLÍČEK, P. & Kovanda, J, (1985): Nové výzkumy kvartéru v okolí Pavlovských vrchů. Sborník Geologických Věd, Antropozoikum 16, 21-59

HARMATA, L. (1987): Late-Glacial and Holocene history of vegetation at Roztoki and Tarnowiec near Jaslo (Jaslo - Sanok Depression). Acta Palaeobot. 27, 43-65

HESS, A. (1968): A trial of reconstruction of the climate in the Holocene in Southern Poland. Folia Quat. 29, 21-39

HÜTTEMANN, H. & BOTRENSCHLAGER, S. (1987): Beiträge zur Vegetationsgeschichte Tirols VI: Riesengebirge, Hohe Tatra - Zillertal, Kühtai. Ber. Naturwiss.-Med. Ver. in Innsbruck 74, 81-112

JANKOVSKÁ, V. (1984): Late Glacial finds of *Pinus cembra* L. in the Lubovnianská kotlina. Folia Geobot. Phytotax. 23, 304-320

JANKOVSKÁ, V. (1988): A reconstruction of the Late-Glacial and Early-Holocene evolution of forest vegetation in the Poprad Basin, Czechoslovakia. Folia Geobot. Phytotax. 23, 304-320

KNEBLOVÁ, V. (1954): Fytopaleontologicky rozbor uhlíků z paleolitického sídlište v Dolních Věstonicích. Anthropozoikum 3, 297-299

KOPEROWA, W. (1958): A Late Glacial pollen diagram at the north foot of the Tatra Mountains. Monographiae Botanicae 7, 107-134

KOPEROWA, W. (1962): The history of the Late Glacial and Holocene vegetation in Nowy Targ Basin. Acta Palaeobot. 2, 3-57

KOPEROWA, W. (1970): Late Glacial and Holocene history of the vegetation of the eastern part of the Jaslo-Sanok Doly (Flysch Carpathians). Acta Palaeobot. 11, 4-42

KRUPIŃSKI, K. M. (1984): Evolution of Late Glacial and Holocene vegetation in the Polish Tatra Mts., based on pollen analysis of sediments of the Przedni Staw Lake. Bull. Pol. Acad. Sci. 31. 38-44

MAMAKOWA, K. & STARKEL, L. (1974): New data about the profil of Young Quaternary deposits at Brzeżnica on the Wisłoka river. Studia Geomorphologica Carpatho-Balcanica 8, 47-59

PAWLIKOWA, B. (1965): Materialy do postglacialnej historii roślinnośći Karpat zachodnich. Folia Quat. 18, 1-9

PLESNÍK, P. (1971): Horná hranica lesa vo Vysokých a Belanských Tatrách. Vydavatelstvo SAV, Bratislava, 238 p.

RALSKA-JASIEWICZOWA, M. (1972): The forest of the Polish Carpathians in the Late Glacial and the Holocene. Studia Geomorphologica Carpatho-Balcanica 6, 5-19

RALSKA-JASIEWICZOWA, M. (1980): Late Glacial and Holocene vegetation of the Bieszczady Mts. (Polish Eastern Carpathians). Panstwowe Wydawnictwo Naukowe, Kraków, 200 p.

RALSKA-JASIEWICZOWA, M. & STARKEL, L. (1975): The leading problems of geography of the Holocene in the Polish Carpathians. Bull. Geol. 19, Warsaw Univ., 27-44

RYBNÍČKOVÁ, E. & RYBNÍČEK, K. (1989): The Holocene development of vegetation in the Oravská kotlina Basin. In: Rybníček, K. (ed.): Excursion Guide Book of the 12th IMEQB, Czechoslovakia, 115-116

RYBNÍČKOVÁ, E. & RYBNÍČEK, K. (1989): Bulhary, the palaeovegetation of the Pavlovian. In: Rybníček, K. (ed.): Excursion Guide Book of the 12th IMEQB, Czechoslovakia, 72-74

ŚRODOŃ, A. (1952): Ostatni glacjał i postglacjał w Karpatach. Biuletyn Państwowego Towarzystwa Geologicznego 63/3, 27-75

ŚRODOŃ, A. (1968): O roślinnośći interstadialu Paudorf w Karpatach zachodnich. Acta Palaeobot. 9/1, 3-28

SZAFER, W. (1952): Schylek pleistocenu w Polsce. Biuletyn Institutu Geologicznego 65, Univ. Warsaw, Warszawa, 33-73

SZAFER, W. (1966): Dziesięc tysięcy lat historii lasu w Tatrach. Nauka dla Wszystkich 1, 3-31

WALTER, H. & STRAKA, H. (1970): Arealkunde - Floristisch-historische Geobotanik. Eugen Ulmer, Stuttgart, 478 p.

Addresses of the authors:

Dr. E. Rybníčková, Institute of Systematic and Ecological Biology, Czechoslovak Academy of Sciences, Květná 8, CS-603 65 Brno, Czechoslovakia
Dr. K. Rybníček, Institute of Systematic and Ecological Biology, Czechoslovak Academy of Sciences, Květná 8, CS-603 65 Brno, Czechoslovakia

The upper timberline dynamics during the last 1100 years in the Polar Ural Mountains

Stepan G. Shiyatov

Summary

Based on studies of the age structure of larch stands growing at their upper limit in the Polar Ural Mountains and on the dendrochronological dating of trunk and root remains of dead trees, a detailed reconstruction of the open larch forests and the upper timberline dynamics during the last 1100 years was made. The role of climatic fluctuations in the alteration of the altitudinal position of the upper timberline and the larch stands structure is discussed.

Résumé

La dynamique des clair-bois du larix et de la limite supérieure forestière pendant les dernières 1100 ans etait reconstruit en detail sur la base de l'étude de la structure de l'âge des peuplements du larix qui poussent à leur limite supérieure dans les montagnes de l'Ural polaire et par le datage des restes des troncs et des racines des arbres morts par l'analyse des cernes. Nous discutons le rôle des fluctuations climatiques dans le changement de la position de la limite supérieure forestière et de la structure des peuplements du larix.

1. Introduction

Usage of direct evidence is of great interest in the studies of the upper timberline dynamics. This evidence comprises primarily the locality of living trees and wood remains of dead trees which have been preserved *in situ* and are found above ground and in various deposits (alluvial, lacustrine, and peat). If we know the location and age of living trees and the calendar time of their appearance and disappearance, we can determine the upper timberline shifts in various time periods with high accuracy.

Numerous explorers (SUKACHEV, 1922; GORODKOV, 1926; SOCHAVA, 1927; ANDREEV, IGOSHINA & LESKOV, 1935) observed a great number of dead trees and wood remains in various degrees of decomposition around the upper timberline on the eastern flank of the Polar Ural Mountains. Such wood is especially abundant in the Sob River Basin. The dating of the wood by the dendrochronological method has shown that some trees died 600-800 years ago (SHIYATOV, 1979). Wood remains of trees are often found above the present timberline. In particular, large areas (up to 192 hectares) of completely dead larch stands were discovered on the east slope of the Rai-Iz Massif. In this area special studies were carried out to reconstruct the upper timberline dynamics during the last millennium and to reveal factors which determine these dynamics.

2. Objects and methods of investigations

The Rai-Iz Massif is situated just to the north of the Polar Circle.The open forests at the upper timberline consist exclusively of Siberian larch (*Larix sibirica* Ldb.). The range of timberline altitudes varies between 280 and 350 m a.s.l. Two large (192 and 39 hectares) and seven small (from 0.2 to 3.8 hectares) completely dead larch stands were found above the present timberline at the watershed of the Kar-Doman-Shor stream. The earlier upper timberline had an altitudinal range between 350 and 410 m a.s.l. There are many standing and lying trunks and roots in various stages of decomposition in these areas (Photo 1) and a large number of dead trees and wood remains in the open larch forests.

To understand the upper timberline dynamics we used data on the age structure (exact age determination of 315 trees), and the regional spread of individual age generations of trees. To determine the life span (in calendar years) of the dead trees, 80 cuts were taken from trunks and roots in various locations. In order to estimate the upper timberline dynamics for the time intervals beyond the maximum age of living trees, a transect 430 m long and 20 m wide was set up across the second largest area (39 hectares) where all trees had died. The transect began at an elevation of 340 m a.s.l. where the highest remains were found, and ended at the present timberline at an elevation of 280 m a.s.l. All remains on the transect were mapped (249 trees). For the tree-ring analysis cuts from trunk and root remains were taken as close to the base of trunks and roots as possible to estimate the time of appearance and dying off of trees with maximum accuracy. On account of the high degree of wood decomposition, cuts could not be taken from 40 remains. Thus, cuts were taken from 209 wood remains only (107 from roots, 102 from trunks). In addition, 16 larch seedlings were found on the transect, their age being up to 80 years and their height up to 4 m. Their locality was also mapped. Ring widths were measured in each cut. Ring widths were plotted and used for relative and absolute dating of rings, for the discovery of missing rings, and for choosing appropriate curves to calculate indices. Based on chronologies obtained from old living trees (the maximum ages of living trees are 350-400 years in this region), it was possible to achieve an absolute dating of the samples taken.

Photo 1 A view of the dead larch stand above the present timberline on the east slope of the Rai-Iz Massif

3. Results of investigations

The structure analysis of present larch stands growing near the upper timberline has shown that they consist of few morphological groups differing in height, diameter, growth form, and crown development. Age determination of trees has shown that these groups also differ in age. In other words, during certain time periods the regeneration was intensive whereas it was weak or lacking during others. Three age generations in the larch stands were distinguished in the area studied:

(1) the over-aged generation which emerged in 1630-1690. At present this generation is represented by a small number of individuals in stands;
(2) the middle-aged generation which emerged in 1780-1850. Presently this generation prevails in larch stands;
(3) the young generation which emerged since the 1920s and is now in the final stage of formation.

Thus, during the last 350 years there were three time periods favourable for regeneration and formation of larch generations and two time periods when regeneration was very weak or lacking. According to KOMIN's classification (KOMIN, 1963), the larch stands at their upper limit of spread may be attributed to cyclic or stepped uneven-in-age distribution.

Using morphological traits one can define which age generation the trees growing at the upper timberline belong to. This enables the estimation of the upper limits of the above-mentioned age generations. It means that we can reconstruct the upper timberline for those time periods when the formation of age generations ceased.

The analysis of the maps shows that the last 350 years saw significant displacements of the upper timberline in the Polar Ural Mountains, especially during the period when the middle-aged generation came into being. This generation occupied the driest and best drained sites. The over-aged generation grew on wet sites, the young generation develops on sites with varying moisture.

As shown in Fig. 1, the most ancient wood was found in the middle part of the transect. The oldest tree-rings were formed in the second half of the ninth century. Probably trees appeared in the lower part of the transect at this time, but their remains have not survived. During the ninth and tenth centuries larch stands were very sparse. The upper limit of isolated trees was at an elevation of 325 m a.s.l. (90 m along the slope from the beginning of the transect) and the upper limit of the light forest was at an elevation of 305 m a.s.l. (210 m from the beginning of the transect).

Intensive regeneration started in the first half of the twelfth century and proceeded until the end of the thirteenth century. The most favourable growing conditions for larch were in the twelfth and thirteenth centuries. Larch stands were most dense during the last millennium. At the beginning and especially in the second half of the eleventh century there was an intensive rise of the upper timberline (from 305 to 340 m a.s.l., corresponding to a relative distance of 190 m along the slope). In the twelfth century the upward movement of the upper timberline slowed down. The maximum was reached in the middle of the thirteenth century (Fig. 1).

At the close of the thirteenth century larch growing conditions deteriorated and comparatively young trees (100-150 years old) began dying off on a large scale. The uppermost trees died out at first.

The most intensive thinning of stands and upper timberline retreat took place towards the close of the fourteenth century. From the end of the thirteenth to the end of the fourteenth centuries the upper timberline sank from 340 to 310 m a.s.l., corresponding to 180 m along the slope.

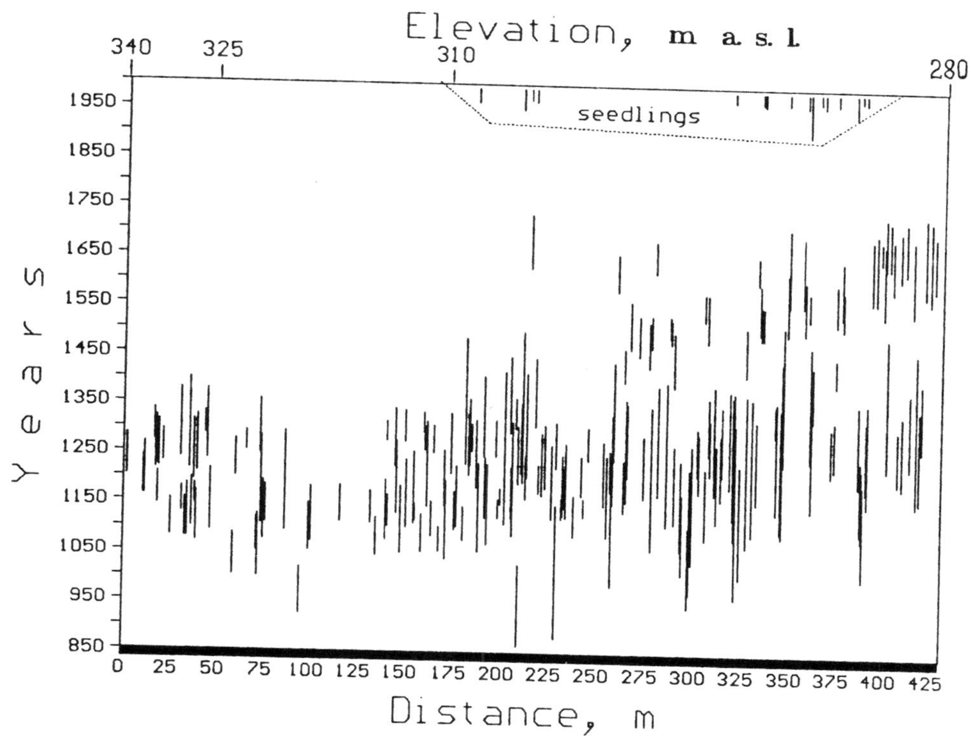

Fig. 1 Distribution of dated wood remains and larch seedlings along the transect

In the first half of the fifteenth century the deterioration of forests came to an end. Moreover comparatively favourable conditions for regeneration were observed in the lower part of the transect.

During the second half of the fifteenth century a further retreat of the upper timberline took place from 310 to 300 m a.s.l.

In the sixteenth century tree growing conditions were again more favourable than in the second half of the fifteenth century. Larch stands increased in density and the upper timberline remained stationary at approximately 300 m a.s.l.

In the first half of the seventeenth century a new retreat of the upper timberline took place (70 m along the slope, corresponding to 300 to 290 m a.s.l.). In the second half of this century the decline of the open larch forests ceased and in the lower part of the transect there was intensive regeneration. During this period the over-aged generation formed, of which some trees are still alive.

In the second half of the eighteenth century all trees growing on the transect died off and the upper timberline retreated to 280 m a.s.l.

In the nineteenth century not one single living tree stood on the transect. The first larch seedlings reappeared at the beginning of the twentieth century at an elevation of about 290 m a.s.l. There are 16 seedlings on the transect now. They are mainly 20-30 years old. In Fig. 1 the seedlings are shown in the upper part of the drawing. Formation of the young larch generation only took place in the lower part of the transect at an elevation of 280 to 310 m a.s.l.

4. Discussion

Thus, during the last 1100 years the upper timberline and the larch stands density altered significantly in the Polar Ural Mountains. From the ninth to the thirteenth centuries larch colonization of the formerly treeless tundra sites took place. The maximum development of open larch forests at their upper limit of spread was observed in the thirteenth century.

During this period the upper timberline moved 35 m upwards, corresponding to a distance of 210 m along the slope. Then forest deterioration and the upper timberline decline began and this process continued to the end of the nineteenth century. From the end of the thirteenth to the end of the nineteenth centuries the retreat of the upper timberline was 60 m in altitude or 430 m along the slope.

The displacements of the upper timberline took place with different intensity in various time periods. However, it can be noted that the advancement of the open larch forests upwards was usually slower than their retreat.

As there was no evidence of forest fires or other catastrophic phenomena in the study area and also no evidence of significant human influences on open larch forests, the forest dynamics were most probably caused only by climatic changes.

To reconstruct the past climatic conditions we used the 1009-year tree-ring chronology of Siberian larch obtained from wood of living trees and wood remains. All wood samples (cuts and cores) were collected in the study area. Of greatest influence upon the tree growth variability was the air temperature of June and July (the correlation coefficient is 0.78). In Fig. 2 one can see the reconstructed air temperature averages for every 20 years of the last millennium (GRAYBILL & SHIYATOV, 1989). During this period summer temperatures fluctuated significantly, their variability range exceeding 1.2°C in various 20-year-intervals.

Two long-term periods can be distinguished: the warm period from the tenth to thirteenth centuries and the period of cooling from the fourteenth to the end of the nineteenth centuries. The warmest period was in the twelfth and thirteenth centuries, the coldest one was in the nineteenth century.

It is not difficult to see that the long-term temperature oscillations coincided very well with the altitudinal changes of the upper timberline. In other words, during the Medieval climatic warming or the Little Climatic Optimum open larch forests at the upper limit of their spread moved to a higher elevation by 60-80 m, compared to their present altitudes. Climatic cooling during the Little Ice Age has resulted in larch stands decline and the upper timberline retreat. There are also short-term fluctuations (of several decades duration) which influenced larch forests dynamics.

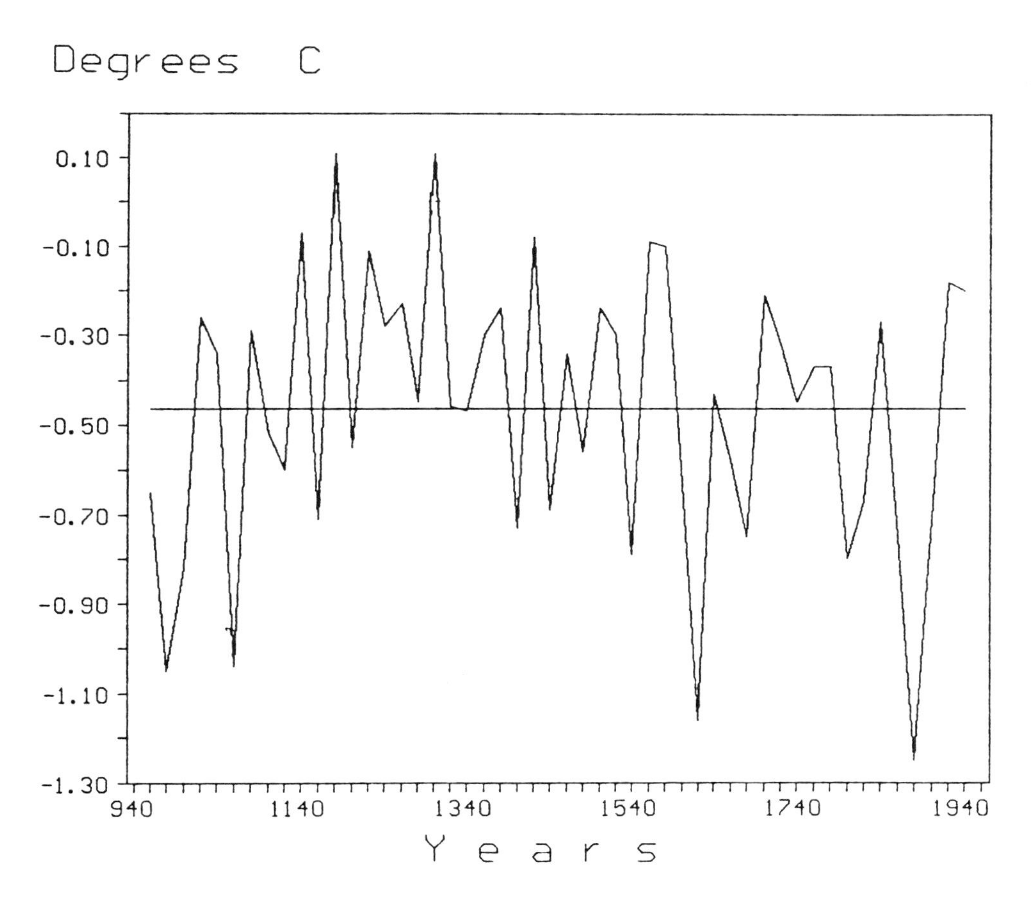

Fig. 2 Twenty year averages of reconstructed average June-July temperature departures (redrawn from GRAYBILL & SHIYATOV, 1989).

The dying off of trees is basically related to cold periods of no less than 20-30 years duration. Such cold periods occurred from the late thirteenth century to the early fourteenth century, in the middle of the fourteenth century, in the first half of the sixteenth century, in the beginning and the end of the eighteenth century. Formation of a single age generation requires favourable climatic conditions during no less than 50-60 years. However, larch seedlings can emerge and grow in cold climatic periods because less heat supply is needed for their growth and development during growing season than for adult trees. Therefore, the appearance and existence of seedlings are not so closely connected with the beginning of the warm period as is the disappearance of trees with cold periods. The alternation of warm and cold periods which is typical of the region causes the formation of cyclic and stepped uneven larch stands. The analysis of the time of appearance and disappearance of trees, from which only trunk and root remains have survived, suggest that in the past the age structure of larch stands was the same as today. We intend to make a further, more detailed analysis of the age structure of dead stands.

Deterioration of forests and lowering of the upper timberline continued up to the beginning of the twentieth century. In connection with the present warming larch seedlings have started to settle in tundra communities since the 1920s. This process was rather intensive and larch seedlings appeared in the lower part of the transect (Fig. 1). The occupation of the tundra by forest would have been more active if more viable seeds had been blown into the treeless areas.

The above-mentioned relation of individual age generations to sites with different moisture conditions is supposed to have been caused by changes in the humidity of climate. The larch stands that were forming from the ninth to the seventeenth centuries are confined to the wettest sites. To all appearance, the climate then was dry and continental.

In the eighteenth and nineteenth centuries, when the middle-aged larch generation was under formation, the humidity increased significantly.

Therefore, during this wet period the advancement of the upper timberline was rather intensive but restricted to the driest slope sites.

References

ANDREEV, V. N.; IGOSHINA, K. N. & LESKOV, A. I. (1935): Reindeer pastures and vegetation cover of the Polar Priuralie (in Russian). Soviet Olenevodstvo 5, 171-406

GORODKOV, B. N. (1926): The Polar Urals in the watershed of the Sob River (in Russian). Proc. Bot. Mus. USSR Acad. Sci. 19, 1-74

GRAYBILL, D. A. & SHIYATOV, S. G. (1989): A 1009 year tree-ring reconstruction of mean June-July temperature deviation in the Polar Urals. In: Noble, R. D.; Martin, J. L. &

Jensen, K. F. (eds.): Air pollution effects on vegetation, including forest ecosystems. Broomall, PA, USA. Department of Agriculture, Forest Service, Northeastern Forest Experiment Station, 37-42

KOMIN, G. E. (1963): Age structure forest stands types (in Russian). Proc. of Institute of Higher Education. For. J. 3, 37-42

SHIYATOV, S. G. (1979): Dating of the last forest expansion in the Polar Ural Mountains (in Russian). In: Biological problems of the North: VIII Symposium. Apatity, 63-64

SOCHAVA, V. B. (1927): Botanical essay of the Polar Ural forests from the Nelkja River to the Khulga River (in Russian). Proc. Bot. Mus. USSR Acad. Sci. 21, 1-71

SUKACHEV, V. (1922): Climate and vegetation changes during the post-Tertiary time in the Siberian north (in Russian). Meteorological Vestnik 1-4, 25-43

Address of the author:

Dr. S. G. Shiyatov, Laboratory of Dendrochronology, Institute of Plant and Animal Ecology, Ural Division of the Russian Academy of Sciences, 8 Marta Street, 202 Ekaterinburg, 620219 GSP-511, Russian Federation

Dendrochronological sampling strategies for radiodensitometric networks in northern hemisphere Subalpine and Boreal zones

Fritz Hans Schweingruber

Summary

Due to systematic sampling strategies and the use of radiodensitometric techniques (X-ray densitometry), the science of dendrochronology is able to establish series of proxy data containing year-by-year information on climatic conditions, especially summer temperatures. Present work includes the construction of a dense network covering climatic conditions in the northern hemisphere over the past 300-400 years and a looser network covering the past 1000-2000 years. A few of these data series may extend over as many as 6000-8000 years. Unfortunately, the infrastructure for the realization of these projects is inadequate.

Zusammenfassung

Mit einer ökologisch geplanten Probenbeschaffung und der Verwendung radiodensitometrischer Techniken (Röntgen-Densitometrie) gelingt es, Chronologien mit jährlichen Informationen über Sommertemperaturen aus Jahrringdichten zu lesen. Es wird am Aufbau eines dichten, sich über die Nordhalbkugel erstreckenden Netzwerkes für die letzten 300 bis 400 Jahre und an einem lockeren Netzwerk für die letzten 1000 bis 2000 Jahre gearbeitet. Auch die Rekonstruktion von Chronologien, die sich über 6000 bis 8000 Jahre erstrecken, wird möglich sein.

1. Introduction

In 1963 POLGE (1963) developed a feasible X-ray technique for dendrochronology (X-ray densitometry). In 1965, FRITTS (1965) constructed the first sampling network for the USA, and in 1971 PARKER & HENNOCH (1971) detected a close correlation between maximum latewood densities in Engelmann spruce at the upper timberline in Canada and the mean monthly August temperature. Based on this discovery, SCHWEINGRUBER et al. (1990) plotted a sampling network for the northern hemisphere and were able to survey some of the regions. The interpretation of the results was based on specially designed technical equipment, stringent selection of sampling sites according to ecological factors, and comparison with sound climatic data. At each stage of interpretation, certain basic factors had to be observed:

X-ray densitometric procedure

The measuring technique has been described by POLGE (1963), PARKER (1971), LENZ et al. (1976), and SCHWEINGRUBER (1988). This technique is far more demanding than simply measuring ring widths and requires not only great technical skill but also durable equipment so that the findings remain comparable and reproducable for years. Of especial importance are careful preparation of the samples by well-trained staff (automated processing has so far not proved useful) and a foolproof system of synchronization through cross-dating.

Site selection

The existing networks are based on the discovery that variations in the maximum density of conifers in different regions can be synchronized with each other (Fig. 1, 2, and 3). While it is easy to synchronize samples from subalpine sites far apart, it seems also possible to synchronize data from high and low sites in close vicinity of each other. In some cases, especially in broadleaf zones, growth seems to be limited by low temperatures in summer, in others rather by lack of precipitation.

The continent-wide networks are arranged so as to obtain maximum information on summer temperatures with the sampling plots located on cool-moist sites which are not affected by summer drought or other factors. In practice this means that the sampling sites are largely located along or near the upper or northern timberlines. For example, cores reflecting summer temperatures were obtained in Western Scotland from sites at only 150 m a.s.l., in the Central Alps from elevations as high as 2200 m. In the Boreal zone of Eurasia, maximum densities in trees on most of the sites are limited by low summer temperatures. Particularly excellent data have been obtained from Scots pines (*Pinus sylvestris*) growing at the timberline of Northern Scandinavia (BRIFFA et al., 1990). Most of the samples from subalpine sites in the Alps and the Carpathian Mountains manifest similar reactions, while in the Mediterranean area only samples from extreme, cool-moist sites very close to the timberline provide usable data. Of the species available, Cembran pine (*Pinus cembra*) has to be excluded because it forms only very narrow latewood rings, and also larch (*Larix decidua*) because its growth rings reflect periodic attack by larch bud moth rather than variations in summer temperature.

The density of the sampling sites within the networks varies, though as a rule they are seldom more than 200-300 km apart; only where no suitable sites can be found are the distances greater. The comparability of the sites is assessed on the basis of phytosociological surveys.

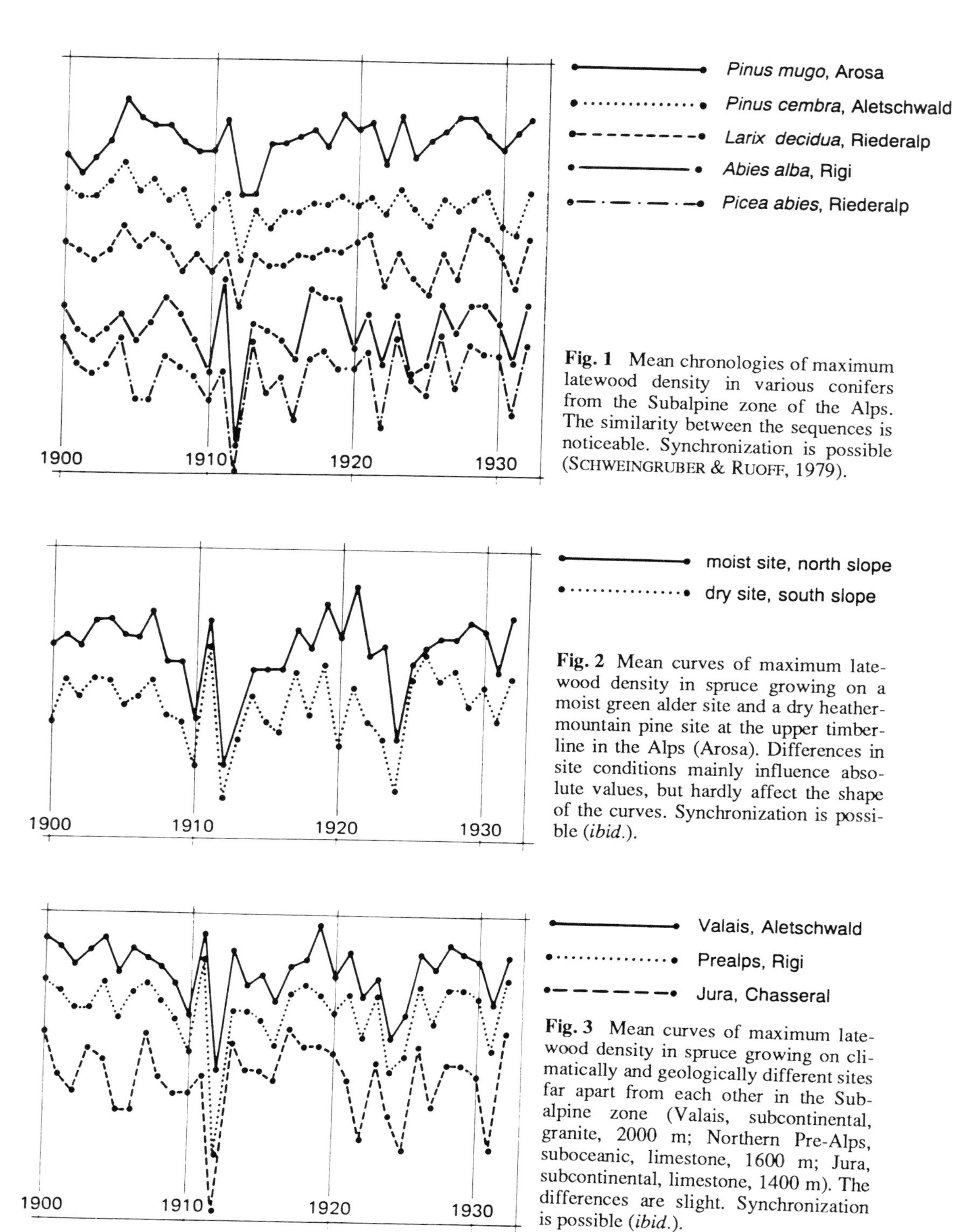

Fig. 1 Mean chronologies of maximum latewood density in various conifers from the Subalpine zone of the Alps. The similarity between the sequences is noticeable. Synchronization is possible (SCHWEINGRUBER & RUOFF, 1979).

Fig. 2 Mean curves of maximum latewood density in spruce growing on a moist green alder site and a dry heather-mountain pine site at the upper timberline in the Alps (Arosa). Differences in site conditions mainly influence absolute values, but hardly affect the shape of the curves. Synchronization is possible (*ibid.*).

Fig. 3 Mean curves of maximum latewood density in spruce growing on climatically and geologically different sites far apart from each other in the Subalpine zone (Valais, subcontinental, granite, 2000 m; Northern Pre-Alps, suboceanic, limestone, 1600 m; Jura, subcontinental, limestone, 1400 m). The differences are slight. Synchronization is possible (*ibid.*).

2. Sample selection

Only cores from trees with "normal" growth furnish comparable data. Trees displaying obvious stress symptoms such as dry terminal shoots, partial decay of crown or roots, buckled or leaning trunks, or, growing on extreme sites, e.g. very shady positions, cliff edges, locations exposed to very strong wind, sites bordering on bogs, were excluded. No cores were taken from parts of trunks displaying obvious tension or compression stress or any other abnormality, e.g. swellings, depressions, wounds. While a long lifespan is desirable, the oldest trees on a given site were often excludedül because they did not meet these requirements; first priority was given to the quality of the material. Two cores were taken from each sample tree, perpendicular to the main axis to facilitate the technical preparation of the samples.

3. Conclusions

Previous studies have shown that:

- continent-wide, year-by-year anomaly maps of summer temperatures over the past 200-400 years can be constructed on the basis of data from recent material, calibrated against recent meteorological records (SCHWEINGRUBER et al., 1990);
- for certain areas it is possible to construct regional dendrochronological sequences covering 1000-2000 years (BRIFFA et al., 1990; SHIYATOV, 1992);
- sequences covering several thousand years can only be constructed for a very few areas;
- the present densitometric and technical infrastructure is of limited capacity, and further laboratories should be set up in Northern and Southern Europe to cover future research needs;
- most of the existing laboratories are not sufficiently acquainted with the various climatological and statistical methods available and in many of them cooperation between research institutes is needed to bring things up to date;
- the costs of dendroclimatological studies covering each of the hemispheres exceed the financial means of the individual laboratories or even national science foundations, but it may be possible to resolve such problems through the ESF.

References

BRIFFA, K. R.; BARTHOLIN, T. S.; ECKSTEIN, D.; JONES, P. D.; KARLÉN, W.; SCHWEINGRUBER, F. H. & ZETTERBERG, P. (1990): A 1400-year tree-ring record of summer temperatures in Fennoscandia. Nature 346, 434-439

FRITTS, H. C. (1965): Tree-ring evidence for climatic changes in Western North America. Monthly Weather Rev. 93, 421-443

LENZ, O.; SCHÄR, E. & SCHWEINGRUBER, F. H. (1976): Methodische Probleme bei der radiographisch-densitometrischen Bestimmung der Dichte und der Jahrringbreiten von Holz. Holzforschung 30, 114-123

PARKER, M. L. (1971): Dendrochronological techniques used by the Geological Survey of Canada. Geol. Surv. Can. Paper 71/25, 1-30

PARKER, M. L. & HENOCH, W. E. S. (1971): The use of Engelman spruce latewood density for dendrochronological purposes. Can. J. For. Res. 1, 90-98

POLGE, H. (1963): L'analyse densitometrique des cliches radiographiques. Ann. Ecole Nat. Eaux et Forêts 20, 531-581

SCHWEINGRUBER, F. H. (1988): Tree rings. Basics and applications of dendrochronology. Dordrecht, Boston, London, 276 S.

SCHWEINGRUBER, F. H. & RUOFF, U. (1979): Stand und Anwendung der Dendrochronologie in der Schweiz. Z. Arch. Kunstgesch. 36, 69-90

SCHWEINGRUBER, F. H.; BRIFFA, K. R. & JONES, P. D. (1991): Yearly maps of summer temperatures in Western Europe from A.D. 1750 to 1975 and Western North America from 1600 to 1982: Results of a dadiodensitrometrical study on tree rings. Vegetation 92; 5-71

SHIYATOV, S. G. (1992): The upper timberline dynamics during the last 1100 years in the polar Ural Mountains. (This volume)

Address of the author

Prof. Dr. F. H. Schweingruber, Swiss Federal Institute for Forest Snow and Landscape Research (WSL), CH-8903 Birmensdorf, Switzerland

Timberlines as indicators of climatic changes: problems and research needs

Friedrich-Karl Holtmeier

Summary

The upper and the polar timberlines are caused by heat deficiency. The timberlines respond to a general cooling with a more or less pronounced delay because of the great inertia of trees once established. Thus, they cannot always be expected to be in balance with existing temperature conditions. The interference of climatic, biotic, and anthropogenic factors represents a further grave problem with regard to the interpretation of previous timberlines. The effects of cyclic mass-outbreaks of pathogenic insects, forest fires and human impact may have altered or even intensified the effects of climate. Altogether former timberlines are less appropriate as indicators of the palaeoclimatic situation than is usually assumed. At the moment it is not possible to make predictions on the potential effects of the present global warming and increasing air pollution on the timberlines.

Zusammenfassung

Die obere und die polare Waldgrenze sind Wärmemangelgrenzen. Wegen des großen Beharrungsvermögens des Waldes reagieren sie nur mit einer mehr oder weniger großen Verzögerung auf Klimaverschlechterungen. Die Waldgrenzen befinden sich daher nicht unbedingt im Gleichgewicht mit den jeweiligen Wärmebedingungen. Ein weiteres gravierendes Problem im Hinblick auf die Interpretation früherer Waldgrenzen ist die Überlagerung der Auswirkungen von klimatischen, biotischen und anthropogenen Faktoren. Die Auswirkungen von Massenvermehrungen von Schadinsekten, Waldbränden und der Eingriffe des Menschen überlagern und verschärfen u. U. die Einflüsse einer Klimaverschlechterung. Waldgrenzen sind daher als Klimazeugen weniger gut geeignet als allgemein angenommen wird. Eine verläßliche Voraussage hinsichtlich der möglichen Folgen der gegenwärtigen Erwärmung und der zunehmenden Luftverschmutzung für die Waldgrenzen ist zur Zeit nicht möglich.

1. Introduction

It is well substantiated and generally accepted that the upper and polar timberlines are caused by heat deficiency. In view of the dependence of the timberlines on warmth, many attempts have been made to detect the critical timberline-controlling temperature (for references cp. HOLTMEIER, 1965, 1974). The locations of the polar timberlines turned out to be related to the average position to the 10°C-isotherm of the warmest month (July). BROCKMANN-JEROSCH (1919) had already shown that this obvious coincidence does not reflect a causal dependence of the timberlines on the isotherm, but only roughly mirrors the controlling role of heat deficiency.

Assuming that the temperature requirements of timberline tree species have not changed appreciably during the Holocene, the former locations of the timberlines could help to understand the history of the Postglacial climate. Nevertheless, there are some general problems which have to be taken into consideration.

2. Problems

2.1 Timberlines: biological boundaries/ecotones

By using previous positions of the upper and polar timberlines as indicators of climatic changes it is easy to begin thinking of the timberlines themselves as "isolines", supposing them to have been in balance with specific temperature conditions. However, timberlines are biological boundaries characterized by great ecological diversity and complexity. They are formed by different tree species, which exhibit distinct ecological properties and demands. Thus, the timberlines do not respond uniformly to their climatic and biotic environment.

There is no good reason to assume that natural ecological conditions were less different or that the variety of timberline types was smaller in the past than at present. Thus, modern timberline ecology may help us to better understand the response of former timberlines to Holocene climatic change. In evaluating timberlines as indicators of climate it appears appropriate to look at actual timberlines which were not influenced, or only moderately influenced, by man. Such timberlines still exist in some parts of the Subarctic and Rocky Mountains.

These timberlines usually occur as ecotones, i.e. transitional belts between the closed high forest and the most advanced solitary trees and tree islands. The widths of these ecotones may range from some tens of meters to many kilometres. The timberline may also occur as a line - at least when seen from afar. On closer examination most "lines" turn out to be ecotones. Thus, when interpreting macro-fossils such as wood fragments, snags, or cones one has to know where they were located within the former ecotone.

The ecotones are characterized by complex ecological dynamics resulting from the influence of the very locally changing site conditions on tree growth, survival rate, and regeneration. Not least the effects of the trees and tree islands themselves on their environment play an important role (HOLTMEIER, 1985, 1986, 1987).

Reproduction by seed occurs only sporadically and may even have been non-existent for many decades. Although trees in the most extreme habitats occasionally bear abundant cone crops they usually do not produce viable seeds. Hence, subfossil cones, seeds, and pollen are not reliable indicators of successful regeneration.

Timberline pines, except for a few very long-lived species such as *Pinus aristata* and *Pinus longaeva*, generally do not live for more than several hundred years. In contrast, species of spruce and firs that produce and propagate by layering (e. g. *Picea engelmannii, Picea abies, Abies lasiocarpa*) are at a great advantage because layering still continues at temperatures that would prevent sexual regeneration (HOLTMEIER, 1986).

They continue to exist and propagate indefinitely provided that they are able to physically resist climatic and biotic injuries, as demonstrated by many studies in the Subarctic and in the Rocky Mountains, where the position of the most advanced tree island is not in equilibrium with the present climate (LARSEN, 1965; TOLMACHEV, 1970; IVES, 1973; NICHOLS, 1974, 1975a, 1975b, 1976; PAYETTE, 1976; PAYETTE & GAGNON, 1979; ELLIOT, 1979; HANSEN-BRISTOW, 1981; HANSEN-BRISTOW & IVES, 1985; IVES & HANSEN-BRISTOW, 1983; HOLTMEIER, 1985, 1986).

Thus, timberlines do not respond to climatic changes as automatically and promptly as the snow line does. Similar conditions are very likely to have also occurred in the Holocene. Therefore, when using the positions of former timberlines as climatic indicators, fluctuations of the palaeoclimate may be easily disguised by the inertia of long-lived trees and conal tree islands in particular.

2.2 Non-climatic factors interfering with climatic effects

2.2.1 Insects, fire

Other factors, such as man, fire, and insects may also have caused timberline declines. Large areas of the mountain birch forest (*Betula tortuosa*) in Northern Finland, for example, were destroyed by the geometrid *Epirrita autumnata* which was cyclically multiplying in huge numbers. After repeated defoliation by the caterpillars of that moth the birches fell victims to the harsh timberline climate and to secondary parasites such as *Hylocoetes dermestioides* and *Agrilus viridis*. On Ailigas Mountain (620 m a.s.l.) near Karigasniemi (Finnish Lapland), for example, an area of about 16 km^2 was lost to alpine tundra after such an infestation in 1927 (NUORTEVA, 1963; Fig. 1). In Northern Utsjoki (Finnish Lapland) wide areas of mountain birch forests were killed by *Epirrita autumnata* in 1965/66. At least 1000 km^2 of the birch forest was expected to be replaced by tundra vegetation (KALLIO & LEHTHONEN, 1973; SEPPÄLA & RASTAS, 1980; HOLTMEIER, 1985).

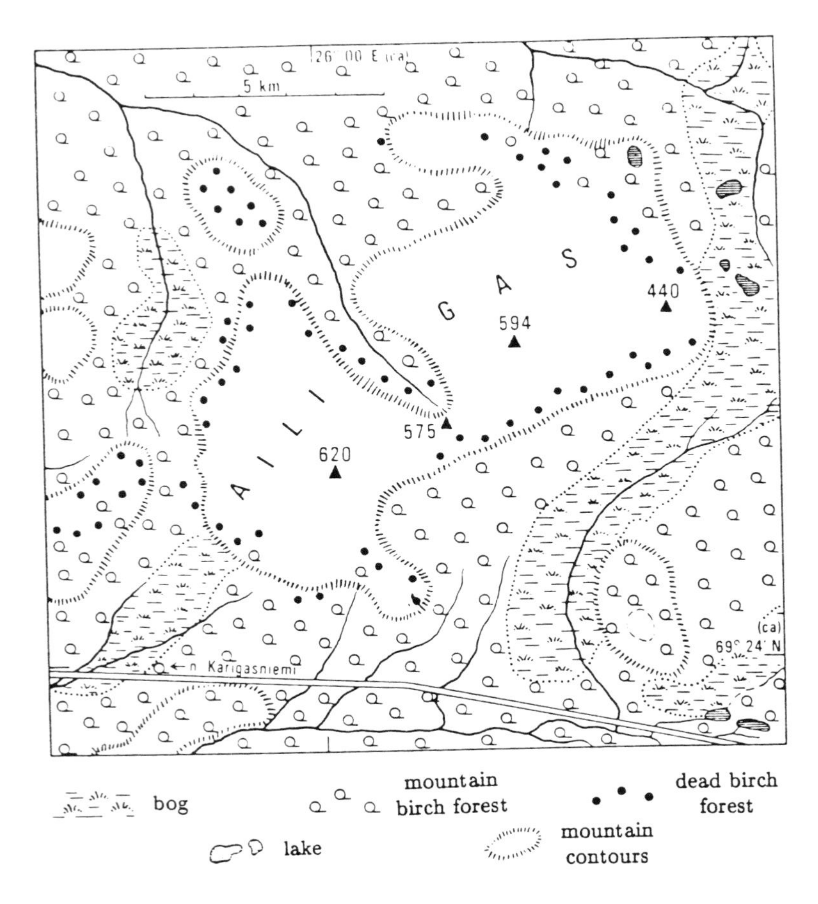

Fig. 1 Depression of the timberline on Ailigas (near Karigasniemi, Northern Finland) by *Epirrita autumnata* (after NUORTEVA, 1963, modified)

Damaged birches usually produce suckers from their root stocks. However, because the fresh stump shoots are heavily browsed by reindeer it takes the birches a long time to recover, if they ever do.

Such events must also be considered in interpreting timberline depressions that occurred prior to the Holocene.

Fire is a further important factor. Forest fires are most effective if they occur during climatically unfavourable periods (cp. BRYSON et al., 1965; NICHOLS, 1975a,b, 1976; PAYETTE & GAGNON, 1979). Under unstable climatic conditions human impact can also trigger timberline decline. After fires and forest break-downs due to insect catastrophes, the growth characteristics of the trees usually change owing to alterations in the nutrient and light conditions.

Thus, varying width and density of tree rings are not necessarily the result of climatic fluctuations (cp. also for KIENAST & SCHWEINGRUBER, 1986)

2.2.2 *Human impact on timberlines*

Timberline history is most difficult to interpret in areas where anthropogenic influences are interfering with natural factors as is the case in the European Alps and in many other high mountain areas of Eurasia which have been settled since prehistoric times.

In the Alps, for example, forest history has been a history of over-utilisation at least since the Middle Ages (cp. HOLTMEIER, 1974, 1986, 1990). The upper timberline was lowered by 150-400 m compared to its uppermost position during the Postglacial Optimum. There is some evidence that the highest Postglacial position of the timberline coincides with the present upper limit of the dwarf shrub *Rhododendron* (EBLIN, 1901; SCHRÖTER, 1908; PALLMANN & HAFFTER, 1933; FURRER, 1957) and podzol soils (PALLMANN & HAFTER, 1933).

However, depressions of the upper timberline were not only caused by human impact but also the result of a climatic deterioration. The effects of this general cooling have been compounded by human influences, at least since the Middle Ages (LAMPADIUS, 1937; ZOLLER, 1967a,b; MAYER, 1970; KÖSTLER & MAYER, 1970; KRAL, 1973; HOLTMEIER, 1974).

At present the climatic limit of tree growth is situated above the actual forest limit as is clearly evidenced in many areas by the invasion of abandoned or rarely used alpine pastures by trees. This invasion was probably encouraged by warming during the second third of our century, as was also the case in many other high mountain areas of the northern hemisphere and at the polar timberline (for references cp. HOLTMEIER, 1974, 1979, 1985, 1986).

In any event tree growth above the closed forest is hampered more by unfavourable site conditions (microclimates, soils, parasitic fungi etc., Fig. 2) than one should expect at the present level of the forest limit. It is obvious that this relatively short positive oscillation of climate could not compensate for the disturbances caused by the removal of the former high elevation forest.

Microclimates and soil conditions changed and the man-induced forest limit has become as pronounced an ecological boundary as the natural timberline had been before. Many sites that were formerly covered by forest became totally unfavourable to natural regeneration and tree growth and there is much evidence that they will remain treeless for a long time (cp. SCHÖNENBERGER, 1985).

That is totally in accordance with the well-known fact that the relationship of temperature to elevation is modified by the mosaic of different microclimates resulting from the influences of the local topography on solar radiation and windflow near the soil surface.

In contrast to widespread opinion the influence of trees and tree clusters on their environment will not necessarily improve the growing conditions, thus making the forest advance in a more or less continuous front.

2.3 Effects of the recent climatic trend

Up to the present, no definite evidence has been reported that would suggest improved tree growth and regeneration at the timberline due to the recent warming trend. On the contrary, many trees which were established during the favourable decades in the middle of our century have already died or become crippled by climatic injuries (cp. HOLTMEIER, 1974; KULLMANN, 1983).

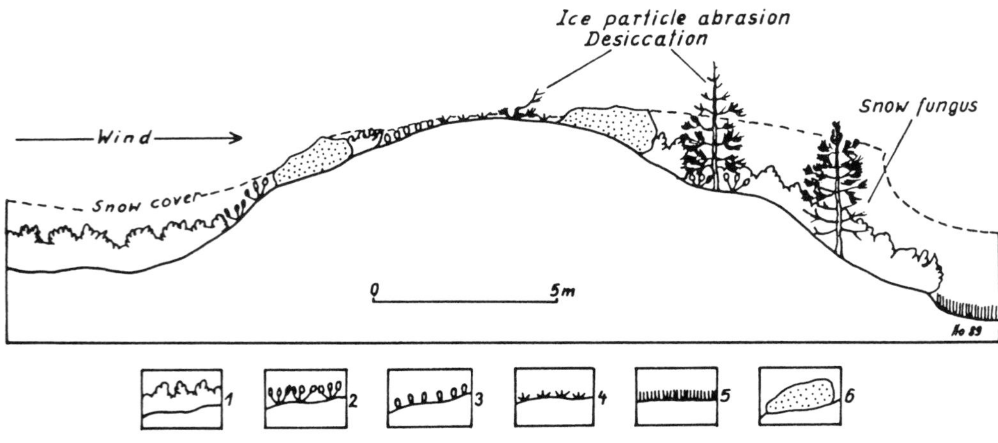

1 Rhododendron ferrugineum; 2 Vaccinium myrtillus; 3 Vaccinium uliginosum, Empetrum nigrum; 4 Loiseleuria procumbens, lichens(Alectoria ochroleuca,Thamnolia vermicularis, Cladonia spec.),Juncus trifidus; 5 Trichophorum caespitosum, Eriopherum scheuchzeri, Carex spec. ; 6 Boulder

Fig. 2 The influence of site conditions on vegetation and natural restocking by *Pinus cembra* above the present man-caused timberline (after HOLTMEIER, 1965, modified)

On the other hand, high-altitude bristlecone pines (*Pinus longaeva, Pinus aristata*) from New Mexico and Colorado showed an increase of radial growth from 1840-1970 which was attributed to rising summer temperatures until about 1960. Continuation of the positive trend after 1960 was explained as a result of increased leaf area related to the preceding warm period. However, more recent data for subalpine limberpines (*Pinus flexilis*) in Nevada and bristlecone pines (*Pinus longaeva*) in the White Mountains of California show that the growth increase persisted through the early 1980s, although no climatic warming trend is reflected by the long-term regional data sets (LA MARCHE et al., 1984; BRADLEY et al., 1982). Consequently the authors concluded that enhanced radial growth must be due to the increasing concentration of atmospheric CO_2. That explanation would be in accordance with TRANQUILLINI (1979), who estimated a decline of 10-20 % in photosynthesic performance of trees at the timberline due to lower natural CO_2 levels.

The interfering effects of both these factors - warming and increasing atmospheric CO_2 - need further investigation, not least with regard to explaining the history of climate and timberlines in the Holocene.

One should be wary of any hasty generalization of these findings because many factors other than increasing CO_2 control tree growth. For instance, we found plant-available phosphorus to be a limiting factor at the upper timberline in the Colorado Front Range (HOLTMEIER & BROLL, 1992).

3. Conclusions and remarks

I leave it at that and come back to the goal of this workshop: i. e. improvement of our knowledge of the Postglacial palaeoclimate by interpreting oscillations of the upper and polar timberlines as indicators of climatic changes and to prepare a basis for perspectives into the future.

Above all we must try to understand the present situation at the timberlines.

There is still a general need for research on the ecological properties and requirements of the different tree species (climatic resistance, susceptibility to pathogenic insects, competition, reproduction etc.).

More detailed local and regional studies on the influences of mid-twentieth century warming on the timberline ecotones (regeneration, survival rates, distribution patterns in relation to site conditions, etc.) would offer a unique opportunity to come to a better understanding of similar events during the Holocene.

However, the actual situation within the ecotones also reflects the site history. In many places the existence of a timberline ecotone is itself the result of previous oscillations of the climate, of infestations by insects, forest fires - and of course anthropogenic influences (cp. HOLTMEIER, 1985).

The site history may be illuminated by means of pollen analysis, tree-ring dating, charcoal dating, analysis of fossil soils, lake sediments, etc. Consequently, there is great need for careful local and regional timberline studies based on a complex approach (Fig. 3; HOLTMEIER, 1974, 1979). These studies should combine careful regional monitoring of present timberlines and ecophysiological research on the tree species represented with palynological investigations, dendro-climatological analysis, radiocarbon dating of macrofossils, and investigations of palaeosoils. Such studies, considering the actual ecological conditions and site history, will enable us to gain the experience we need to interpret former timberlines and to discover the regional differences in the Holocene climate development.

Prediction of the future climate and its potential influences on timberline constitutes a further problem. Although the global warming cannot be contested any more, regional differences are obvious (JONES & WIGLEY, 1990). However, at present no mathematical model is able to answer precisely the question of how regional climates will change (FLOHN, 1978).

Natural rapid changes such as the sudden cooling that occurred during the Little Ice Age and its forerunners cannot be excluded (FLOHN, 1979). Volcanic eruptions that influence the atmosphere, as did the eruptions of Krakatau and Katmai for example, cannot be forecast. Moreover, the effects of future extreme events (droughts etc.) cannot be calculated. Most of them would be of regional importance only. However, it is their prediction that would be of primary interest.

Any extrapolation into the future of oscillation cycles such as SIREN (1963) did, based on his studies on tree growth in the European Subarctic since the twelfth century, must remain pure speculation. He predicted a general decrease of summer temperature in Northern Europe and expected summer temperature to remain low during the twenty-first century.

Also, the simulations of KAUPPI & POSCH (1988) on the potential effects of CO_2-induced warming (2 x CO_2 scenario) on the productivity of the boreal forests and the extension of the boreal zone cannot give any useful answer since only one factor is considered. In view of the complexity of the forest ecosystems it does not make sense to leave out the "fertilizing effect" of increased CO_2 (cp. LA MARCHE et al., 1984) or, for example, of excessive input of atmospheric nitrogen.

Excessive nitrogen is hypothesized as impeding the cold-hardening process and as changing the allocation pattern of carbon and nutrients within the plants. Consequently, the trees will be predisposed to cold damage and secondary damage by fungi, insects, and nematodes.

Scientific Approach and Research Needs

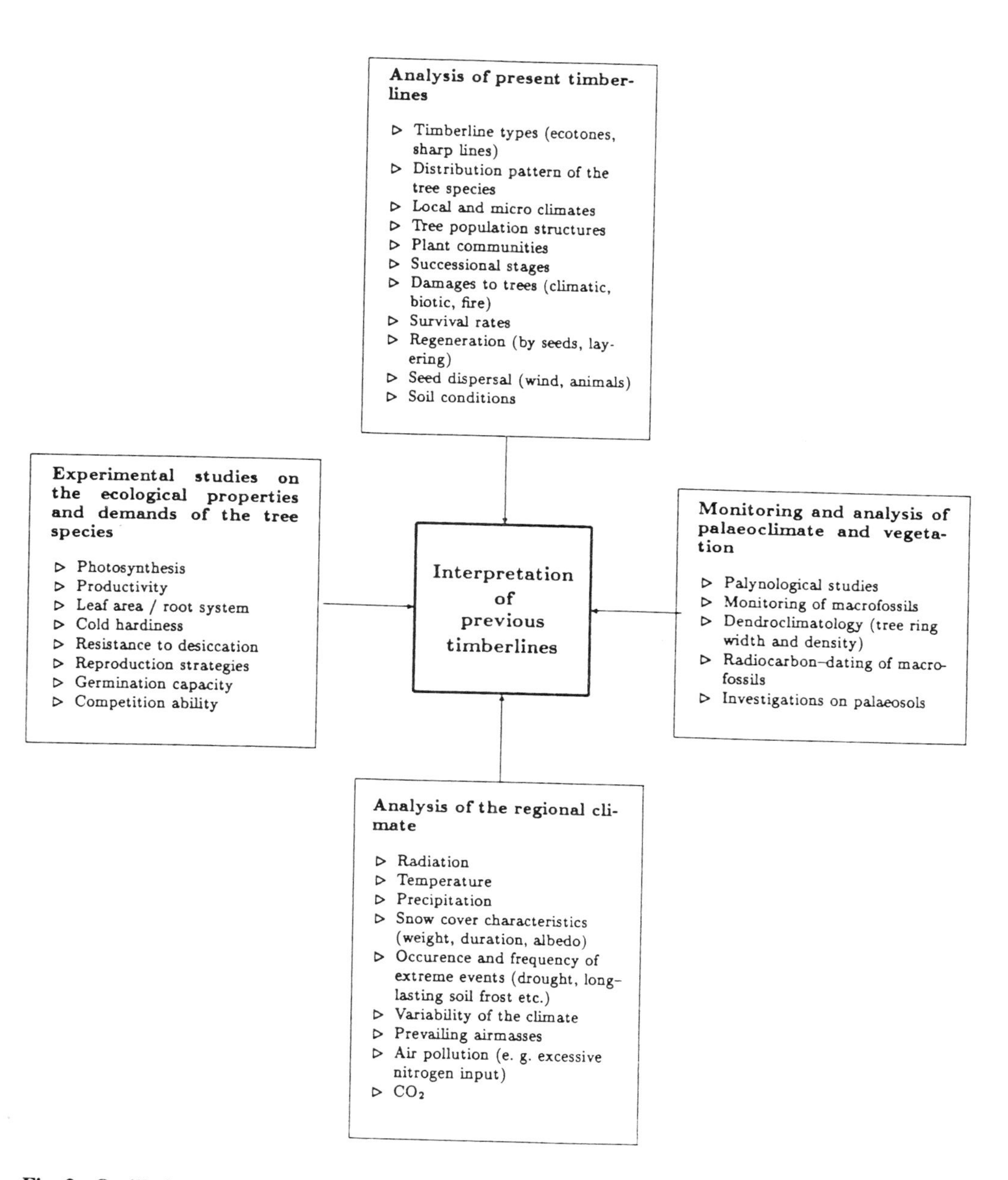

Fig. 3 Oscillations of the alpine and polar timberlines as indicators of climatic changes during the Holocene

Because of the increased growth activity other nutrients such as phosphorus may become the decisive limiting factor. In addition, the input of acid depositions must be considered. Thus, magnesium and calcium are leached out from needles and soils and may also become limiting factors.

These hypotheses (cp. HINRICHSEN, 1986) need to be examined at the upper and polar timberlines before developing even simple simulation models.

Altogether, because the effects of different climatic factors and nutrients are superimposed, and because of the heterogeneous nature and different local history of timberlines, any reliable forecast appears to be impossible.

References

BRADLEY, R. S.; BARRY R. G. & KILADIS, G. (1982): Climatic fluctuations of the Western United States during the period of instrumental records. Contribution 42, Dept. of Geology and Geography, Univ. of Massachusetts, Amherst

BROCKMANN-JEROSCH, H. (1919): Baumgrenze und Klimacharakter. Beitr. Geobot. Landesaufn. 6, 255

BRYSON, R. A.; IRWING, W. N. & LARSEN, J. A. (1965): Radiocarbon and soil evidence of former forest in the Southern Canadian tundra. Science 147, 46-48

EBLIN, B. (1901): Die Vegetationsgrenzen der Alpenrosen als unmittelbare Anhalte zur Festsetzung früherer bzw. möglicher Waldgrenzen in den Alpen. Schweiz. Z. Forstw. 52, 133-157

ELLIOT, D. (1979): The current regenerative capacity of the Northern Canadian trees, Keewatin, N. W. T., Canada: Some preliminary observations. Arct. Alp. Res. II/2, 243-251

FLOHN, H. (1979): Possible climatic consequences of a man-made global warming. Int. Institute for Applied System Analysis, Laxenburg

FLOHN, H. (1989): Ändert sich unser KLima? Neue Erkenntnisse und Folgerungen. Mannheimer Forum 88/89, 135-189

FURRER, E. (1957): Das schweizerische Arvenareal in pflanzengeographischer und forstgeschichtlicher Sicht. Ber. Geobot. Forschungsinst. Rübel in Zürich für 1956, 16-23

HANSEN-BRISTOW, K. J. (1981): Environmental controls influencing the altitude and form of the forest-alpine tundra ecotone, Colorado Front Range. Ph. D. thesis, Univ. of Colorado, Boulder

HANSEN-BRISTOW, K. J. & IVES, J. D. (1985): Composition, form and distribution of forest-alpine tundra ecotone, Indian Peaks, Colorado, U.S.A. Erdk. 39, 286-295

HINRICHSEN, D. (1986): Multiple pollutants and forest decline. Ambio 15, 258-265

HOLTMEIER, F. K. (1965): Die Waldgrenze im Oberengadin in ihrer physiognomischen und ökologischen Differenzierung. Diss. Bonn

HOLTMEIER, F. K. (1974): Geoökologische Beobachtungen und Studien an der subarktischen und alpinen Waldgrenze in vergleichender Sicht. Erdwiss. Forsch. VIII, 130

HOLTMEIER, F. K. (1985): Die klimatische Waldgrenze - Linie oder Übergangssaum (Ökoton)? Ein Diskussionsbeitrag unter besonderer Berücksichtigung der Waldgrenzen in den mittleren und hohen Breiten der Nordhalbkugel. Erdkunde 39, 271-285

HOLTMEIER, F. K. (1986): Über Bauminseln (Kollektive) an der klimatischen Waldgrenze unter besonderer Berücksichtigung von Beobachtungen in verschiedenen Hochgebirgen Nordamerikas. Wetter und Leben 38/3, 121-139

HOLTMEIER, F. K. (1987): Der Baumwuchs als klimaökologischer Faktor an der oberen Waldgrenze. Münstersche Geogr. Arb. 27, 145-151

HOLTMEIER, F. K. (1990): Disturbance and management problems in larch-cembra pine forests in Europe. Proceedings - Symp. on Whitebark Pine Ecosystems: Ecology and management of a high-mountain resource. USDA-Forest Service, Intermountain Res. Station, General Techn. Rep. INT-270, 25-36

HOLTMEIER, F. K. & BROLL, G. (1992): The influence of tree islands and microtopography on pedoecological conditions in the forest-alpine tundra ecotone on Niwot Ridge, Colorado Front Range. Arct. Alp. Res. 24/3

IVES, J. D. (1973): Studies in high altitude geoecology of the Colorado Front Range. A review of the research program of the Institute of Arctic and Alpine Research. Arct. Alp. Res. 5/3, Pt. 2, 67-75

IVES, J. D. & HANSEN-BRISTOW, K. J. (1983): Stability and instability of natural and modified upper timberline landscapes in the Colorado Rocky Mountains, U.S.A. Mount. Res. Develop. 3/2, 149-155

JONES, P. D. & WIGLEY, T. M. (1990): Die Erwärmung der Erde seit 1850. Spektr. Wiss., Okt., 108-166

KALLIO, P. & LEHTHONEN, J. (1973): Birch forest damage caused by *Oporinia autumnata* (BKH.) in 1965-1966 in Utsjoki, N-Finland. Rep. Kevo Subarctic Res. Stat. 10, 55-59

KAUPPI, P. & POSCH, M. (1988): A case study of the effects of CO_2-induced climate warming on forest growth and the forest sector: (A) productivity reactions of northern boreal forests. In: Parry, M. L.; Carter, T. R. & Konijin, N. T. (eds.): The impact of climatic variations on agriculture, Vol. I. 183-195

KIENAST, F. & SCHWEINGRUBER, F. H. (1986): Dendroecological studies in the Front Range, Colorado, U.S.A. Arct. Alp. Res. 18/3, 277-288

KÖSTLER, N. & MAYER, H. (1970): Waldgrenzen im Berchtesgadener Land. Jubiläumsjahrbuch 1900-1970, Ver. Schutze Alpenpfl. Tiere, 1-33

KRAL, F. (1973): Zur Waldgrenzdynamik im Dachsteingebiet. Jb. Ver. Schutze Alpenpfl. Tiere 38, 71-79

KULLMAN, L. (1983): Past and present tree-lines of different species in the Handölan valley, Central Sweden. Coll. Nordicana 47, 25-45

LA MARCHE, V. C.; GRAYBILL, D. A.; FRITTS, H. C. & ROSE, M. R. (1984): Increasing atmospheric carbon dioxide: Treering evidence of growth enhancement in natural vegetation. Science 255, 1019-1021

LAMPADIUS, G. (1937): Die Höhengrenzen der Cima d'Asta und des Lagorai Gebirges. Berliner Geogr. Arb. 15

LARSEN, J. A. (1965): The vegetation of the Ennadi Lake area, N.W.T.: Studies in subarctic and arctic bioclimatology. Ecological Monographs 35 /1, 37-59

MAYER, H. (1970): Waldgrenzen in den Berchtesgardener Kalkalpen. Mitt. ostalpin-din. Ges. Vegetationskd. 11, 109-120

NICHOLS, H. (1974): Arctic North America palaeoecology: the recent history of vegetation and climate dedued from pollen analysis. In: Ives, J. D. & Barry, R. G. (eds.): Arctic and Alpine Environment

NICHOLS, H. (1975 a): The time perspective in northern ecology: Palynology and the history of Canadian boreal forest. Proc. Circumpolar Conf. in Northern Ecology, Sept. 15-18, I157-I165

NICHOLS, H. (1975 b): Palynological and palaeoclimatic study of the Late Quaternary displacement of the boreal forest-tundra ecotone in Keewatin & Mackenzie, N.W.T., Canada. Inst. of Arctic and Alpine Res., Univ. of Colorado, Occ. Paper RM-200

NICHOLS, H. (1976): Historical aspects of the Northern Canadian treeline. Arctic 29/1, 38-47

NUORTEVA, P. (1963): The influence of *Oporinia autumnata* (BKH.) (Lep., Geometridae) on the timberline in subarctic conditions. Ann. Ent. Fenn. 29, 270-277

PALLMANN, H. & HAFFTER, P. (1933): Pflanzensoziologische und bodenkundliche Untersuchungen im Oberengadin. Ber. Schweiz. Bot. Ges. 42, 357-466

PAYETTE, S. (1976): Sucession écologiques des forets d'épinette blanche et fluctuations climatiques. Poste de la Baleine, Nouveau Quebec. Canad. J. Bot. 54, 1394-1402

PAYETTE, S. & GAGNON, R. (1979): Tree-line dynamics in Ungava Peninsula, Northern Quebec. Holarctic Ecology 2, 239-248

SCHÖNENBERGER, W. (1985): Performance of a high altitude afforestation under various site conditions. Proc. 3rd IUFRO workshop, Eidg. Anst. Forstl. Versuchsw. Ber. 270, 233-240

SCHRÖTER, C. (1908): Das Pflanzenleben der Alpen. Zürich

SEPPÄLA, M. & RASTAS, J. (1980): Vegetation map of northernmost Finland with special reference to subarctic forest limits and natural hazards. Fennia 158, 41-61

SIREN, G. (1963): Tree rings and climate forecast. New Scientist 4, 20

TOLMACHEV, A. (1970): Die Erforschung einer entfernten "Waldinsel" in der Großlandtundra. Coll. Geographicum 12, 98-103

TRANQUILLINI, W. (1979): Physiological ecology of the alpine timberline. Ecological Studies 31

ZOLLER, H. (1967 a): Postglaziale Klimaschwankungen und ihr Einfluß auf die Waldentwicklung Mitteleuropas einschließlich der Alpen. Ber. Dt. Bot. Ges. 80/10, 690-696

ZOLLER, H. (1967 b): Holocene fluctuations of cold climate in the Swiss Alps. Rev. Palaeobot. Palynol. 2, 267-269

Address of the author:

Prof. Dr. F.-K. Holtmeier, Institut für Geographie, Abteilung Landschaftsökologie, Westfälische Wilhelms-Universität Münster, Robert-Koch-Straße 26, D-4400 Münster, F.R.G.

Recommendations of Working Group I on "Physiology, ecology, and meteorology at tree limits in Europe"

State of the art

Tree limits are tension zones (ecotones) of great ecological complexity. They vary geographically with respect to structure and function in consequence of both contemporaneous and historical ecological factors.

For few regions, mainly in the Scandes and in the Alps, we have long-term studies of timberline climate and ecophysiological parameters which give a good picture of the relationship between climate and tree limits.

The timberline and tree limits in Europe (the region in question) consist of various types. There are considerable differences between these various types: scattered forests, closed forests, patchy stands and the different species at the tree limits imply different ecological situations so that concepts of one of these types cannot simply be transferred to other types of timberlines and tree limits.

Available data

The results of few long-term studies in some places in the Alps and in Northern Scandinavia are available. They are mainly based on forest research projects concerning forest-economical relevant species and regions. These studies cover only a few of the various types of tree limits. The causal relationships between changes in climate and migrations of the timberline are assessable only on a very regional level.

Research priorities

Systematics of timberlines

To improve the comprehension of the timberline concept in Europe, more investigations concerning the systematics of the various types of tree limits are needed. Future research work has to take these geographical differences generally into consideration. An important aspect is to find sites with primeval tree limits, since the main focus must be the assessment of the natural tree limit processes.

Influence of climatic changes on tree limits

Monitoring of the altitudinal position of the tree limit for the time slices chosen within the EPC programme is highly recommended. Old documentary records of tree limits are invaluable and should be used as baseline for monitoring networks. Reconstructions of palaeoclimate by migrations of the alpine and polar tree limits need a clear picture of the present conditions to calibrate the results of pollen analysis and dendrochronology. In order to better understand the correlations between climatic changes and migrations of the timberlines long-term monitoring studies are urgently needed. They have to take into consideration the various types of timberlines and should concentrate on the influence of the different climatic factors.

Quantification of climatic influence on timberlines

Close cooperation between meteorologists and biologists should lead to an extensive concept of climatic conditions near tree limits. However, the influence of different climatic parameters on the migrations of timberlines is still impossible to quantify. Future studies have to distinguish between macro-climatic and eco-climatic influences to enable reconstructions of past European climates on a macro-climatic and on a regional scale.

Lacks of knowledge

Physiology

Root zone physiology and allelopathy are fields which deserve increased scientific attention. The relevance of root zone conditions and physical dynamics for the occurrence of dwarfs through losses of roots and the stresses related to low soil temperatures need to be investigated in greater detail. It is further suggested that physiological studies also include the seedling/sapling stages.

Population structure

Demographic processes of predominant trees and field-layer species are subjects of great importance. Aspects to be studied are recruitment, mortality, vitality, and effects of herbivores. The response time to various climatic oscillations is fundamental, as well as possible inertial mechanisms, particularly of closed tree populations at the tree limit. Other important issues are the influence of extreme climatic (and geomorphological) events *versus* smooth changes of average conditions.

Ecosystem level

The quantification of influences of various climatic parameters on the migration of tree limits is not possible at present. More substantiated concepts of the mechanisms of climatic influence on tree limits are needed. The importance of extreme events compared to changes in average conditions is not yet known, so improved studies on micro-climatic conditions with special emphasis on radiation and precipitation (snow cover characteristics in particular) in cooperation with meteorologists are recommended.

General recommendations

The need for long-term monitoring studies on the physiological and population ecological levels is stressed. Such studies, on a broad geographical scale in Europe, are important in order to obtain links between the present and the past. On this basis, past tree limits may provide palaeoclimatic information of general validity.

Future workshops organized within the EPC programme and dealing with past climates and timberline migration should strongly focus on the regional aspects, integrate the results of regional research, and stimulate future investigations within the EPC programme.

Group members:

S. Eggenberg, Botanisches Institut, Universität Bern, Altenbergrain 21, CH-3013 Bern, Switzerland
Dr. M. Groß, Botanisches Institut, Universität Hohenheim, Garbenstraße 30, D-7000 Stuttgart 70, F.R.G.
Dr. B. Holmgren, Abisko Research Station, S-98024 Abisko, Sweden
Prof. Dr. F.-K. Holtmeier, Institut für Geographie, Abteilung Landschaftsökologie, Westfälische Wilhelms-Universität Münster, Robert-Koch-Straße 26, D-4400 Münster, F.R.G.
Dr. L. Kullman, Department of Physical Geography, University of Umeå, S-901 87 Umeå, Sweden
Prof. Dr. W. Tranquillini, Institut für Botanik, Universität Innsbruck, Sternwartestraße 15, A-6020 Innsbruck, Austria
Prof. Dr. B. Sveinbjörnsson, University of Alaska, Anchorage, 3211 Providence Drive, Anchorage AK 99508, U.S.A.

Recommendations of Working Group II on "Tree-ring / climate / timberline relations"

General discussion

The group stressed that dendroclimatology remains a largely empirical science. The potential for extracting climate information from tree-growth data has been well established for many parts of Europe. However, the growth processes are complex and there remains considerable scope for experimenting with different tree-growth and climate variables. There is a need for much more work on the ecophysiology of trees of many species and in many areas to help guide and interpret the statistically derived tree-growth / climate relationships.

In Europe, in alpine and high-latitude regions, climate reconstruction has been largely focused on simple temperature variables. Variations in the strength and seasonal response of different tree-ring variables suggests scope for expanding and improving the reconstructions. With the exception of the Mediterranean, there has been less European work reconstructing precipitation-related parameters. More work is needed. The group felt that, until better control can be exerted over the dating and replication of other potentially high-resolution data such as lake-sediment data, it will be difficult to establish reliable cross correlations between them and tree-ring data.

There are potential limitations on the amount of long-timescale climate variability that is recoverable from tree-ring series. The limitations are dependent on the age of individual tree series and on the standardization techniques used to remove non-climate-related variance. The quality of reconstructions is therefore dependent on, but also limited by, the statistical techniques used to produce them. The group felt that further work should address the problems of extracting the maximum reliable long-timescale information from tree-ring series.

The group stressed the importance of collaborative projects where data collection, material and data processing, climatological, ecological and statistical expertise, could be brought together.

There is also a need to expand the technical capacity for undertaking further densitrometric work in Europe (at present only one laboratory is producing data on a routine basis).

The ESF should consider some contribution to the establishment of new densitrometric facilities. Possible centres for these should include the former Soviet Union, Scandinavia and the Mediterranean region.

The group made the following specific recommendations:

(1) Work on the ecophysiology of trees at high altitudes and high latitudes should be continued:

    (a) investigations should not be restricted to the tree line itself but include studies of sub-marginal areas where the potential for building long tree-ring series is greater;
    (b) a range of tree-ring-related parameters should be examined:

        - different ring-width variables;
        - densitrometric variables;
        - isotopic data;
        - ring-morphological features;
        - tree-ring chemistry.

(2) The group felt that a funding priority should be the expansion of the sampling networks into Eastern Europe and the former Soviet Union.

(3) The established potential for climate reconstruction using high latitude and high altitude conifers justifies support for developing the following in these areas:

(a) dense network of ~300-year-old chronologies (including density);
(b) selected smaller network of more regional representative ~1000-year-old chronologies;
(c) a small number of ~6000-8000-year chronologies.

Group members:

Dr. Th. Bartholin, Department of Quaternary Geology, University of Lund, Tornavägen 13, S-223 63 Lund, Sweden
Dr. K. Briffa, Climatic Research Unit, University of East Anglia, Norwich NR4 7TJ, United Kingdom
Dr. M. Eronen, University of Joensuu, Karelian Institute, P.O. Box 111, SF-80101 Joensuu, Finland
Dr. K. Nicolussi, Institut für Botanik, Universität Innsbruck, Sternwartestraße 15, A-6020 Innsbruck, Austria
Prof. Dr. F. H. Schweingruber, Swiss Federal Institute for Forest Snow and Landscape Research (WSL), CH-8903 Birmensdorf, Switzerland
Dr. S. G. Shiyatov, Laboratory of Dendrochronology, Institute of Plant and Animal Ecology, Ural Division of the Russian Academy of Sciences, 8 Marta Street, 202 Ekaterinburg, 620219 GSP-511, Russian Federation

Recommendations of Working Group III on "Palaeo-timberlines and stratigraphic proxy data"

The group discussion was focussed around four key questions:

(a) What types of evidence are available now?
(b) What does each tell us about climate?
(c) How can the different types of evidence be integrated?
(d) What do we need to do in the future?

Our conclusions and recommendations are related to these questions.

(1) A wide range of sources of evidence exists, including various types of peat, megafossiles (particularly tree trunks), lake sediments, soils, moraines and till sheets, glaciofluvial sequences and periglacial deposits.

(2) There is an even wider range of methods that can be used for palaeo-reconstruction. These include the use of pollen, other micro-fossils, macro-fossils, phytoliths, charcoal, beetles, and other animal remains, stable isotopes, and glaciological and sedimentological techniques.

(3) The strengths and weaknesses of these sources and methods are quite well known. For example, the main limitation of pollen analysis is its imprecision, particularly with respect to identification to species level. However, pollen influx records offer greater possibilities than relative pollen counts for the reconstruction of timberlines and palaeoclimate.

(4) Because the different approaches are complementary there should be continued research on a wide front with as many different and interdisciplinary approaches as possible being integrated. Some approaches are more suitable for particular areas, and the strengths of one technique may cover limitations of another. As a starting point, the wealth of information stored in the international literature may be exploited.

(5) Certain established approaches are particularly promising and therefore should be given high priority by national grant-awarding bodies:

   (I) studies of high-resolution pollen influx records and associated research on modern pollen rain;
   (II) detailed studies of plant macro-fossil records from a much larger number of sites than is presently available (the potential of plant macro-fossils has been seriously under-exploited);
   (III) continued studies of plant megafossils above present tree limits, particularly where complete chronologies can be produced for the whole or major parts of the Holocene;
   (IV) intensive studies of glacier variations from glacial lake sediments (including varved sediments) which are potentially able to provide a continuous Holocene record;
   (V) basic research on alternative "physical" proxy data sources, e.g. periglacial deposits and soils.

(6) Despite the large number of sources and methods that have yielded large quantities of data, insufficient attention has been given to calibration of data in terms of climate. Relatively few approaches can provide unambiguous, quantitative climatic reconstructions. There is a special need for the reconstruction of accurate precipitation estimates to match those available or being developed for temperature.

(7) There are also relatively few approaches capable of providing high-resolution records. Annually-laminated sediments (such as varves) have been under-exploited in this regard.

(8) In addition to the development of these established sources and methods, every encouragement should be given to the development of novel techniques in this field. Support for established research should not be allowed to prevent support being given to the testing out of new techniques and new ideas, both of which are needed.

(9) Integrated, interdisciplinary research projects are needed in regions where there is the possibility of establishing several independent, high-resolution records of Holocene climatic variations from multiple data sources. At present such an integration is close to being achieved only in the Alps, where near-continuous records are already available of Holocene variations in summer temperature, equilibrium line, and tree limits. There is great potential in Scandinavia for this kind of research.

(10) It is probably too early to produce regional syntheses for more than a few areas of Europe. This is, however, partly due to the limited exchange of information and ideas between the different workshops like this one; the ESF should consider hosting "regional meetings". This could include field meetings at particularly important sites where an interdisciplinary approach involving scientists from other parts of Europe would be likely to produce further advances.

(11) Although one of the aims of the EPC programme is to reconstruct time slices for selected intervals of Holocene time in order to enable a sound climate modelling, it is this group's firm opinion that this must be parallelled with the development of continuous, high-resolution records for the whole of the Holocene.

(12) Progress in the field of climatic reconstruction is intimately linked with obtaining good dating control. It is important that research should exploit approaches and sites where accurate dating is possible. This field of research relies heavily on radiocarbon dating, cross-checked by dendrochronology.

(13) There is a need for an increase in the capacity of and access to radiocarbon dating laboratories, particularly in relation to AMS dating. In view of the need for precise dating with small samples on the Holocene timescale, the ESF should consider recommending that national scientific organizations expand their support for conventional radiocarbon dating and provide subsidies for AMS dating.

(14) The group felt there was a special need for cooperation with the countries of Eastern Europe. This extends not only to information exchange but also to the exchange of expertise. Furthermore, the countries of Eastern Europe have a shortage of equipment and facilities in some fields (such as AMS dating). One possible way of providing support would be for the ESF to host a "regional meeting" in the Carpathians, to the mutual benefit of all countries involved.

Group members:

Prof. Dr. B. Ammann, Botanische Institute und Botanischer Garten, Universität Bern, Altenbergrain 21, CH-3013 Bern, Switzerland
Prof. Dr. S. Bortenschlager, Institut für Botanik, Universität Innsbruck, Sternwartestraße 15, A-6020 Innsbruck, Austria
Dr. S. Hicks, Department of Geology, University of Oulu, Linnanmaa, SF-90570 Oulu, Finland

Dr. P. Huttunen, University of Joensuu, Karelian Institute, P.O. Box 111, SF-80101 Joensuu, Finland
Dr. H. Hyvärinen, Department of Geology, Division of Geology and Palaeontology, University of Helsinki, Snellmaninkatu 5, SF-00170 Helsinki, Finland
Prof. Dr. W. Karlén, Department of Physical Geography, Stockholm University, S-106 91 Stockholm, Sweden
Dr. M. Kvamme, Botanical Institute, University of Bergen, Allégaten 41, N-5007 Bergen, Norway
Dr. J. A. Matthews, Department of Geology, Univ. of Wales College of Cardiff (UWCC), P.O. Box 914, Cardiff CF1 3YE, Wales, United Kingdom
Dr. E. Rybníčková, Institute of Systematic and Ecological Biology, Czechoslovak Academy of Sciences, Květná 8, CS-603 65 Brno, Czechoslovakia
Prof. Dr. K.-D. Vorren, University of Tromsø, N-9000 Tromsø, Norway

PERIODICAL TITLE ABBREVIATIONS

Abh. Akad. Wiss.	• Abhandlungen der Akademie der Wissenschaften
Abh. Westf. Mus. Naturk.	• Abhandlungen des Westfälischen Museums für Naturkunde
Acta Lapp. Fenn.	• Acta Lapponica Fenniae
Acta Bot. Fenn.	• Acta Botanica Fennica
Acta Palaeobot.	• Acta Palaeobotanica
Acta Phytogeogr. Suec.	• Acta Phytogeographica Suecica
AmS	• Archaeological Museum of Stavanger
Ann. Acad. Sci. Fenn.	• Annales Academiae Scientiarum Fennicae
Ann. Bot. Fenn.	• Annales Botanici Fennici
Ann. Bot. Soc. Zool. Bot. Fenn.	• Annales Botanici Societatis Zoologicae Botanicae Fennicae
Ann. Ecole Nat. Eaux et Forêts	• Annales d'Ecole Nationale des Eaux et Forêts
Ann. Ent. Fenn.	• Annales Entomologici Fennici
Ann. Glaciol.	• Annals of Glaciology
Årb. Univ. Bergen, Mat.-Naturv. Serie	• Årbok fuer Universität i Bergen, Matematisk-Naturvitenskapelig Serie
Arct. Alp. Res.	• Arctic and Alpine Research
Aust. J. Plant Physiol.	• Australian Journal of Plant Physiology
Beitr. Geobot. Landesaufn.	• Beiträge zur Geobotanischen Landesaufnahme
Ber.	• Berichte
Ber. Dt. Bot. Ges.	• Berichte der Deutschen Botanischen Gesellschaft
Ber. Geobot. Inst., Stiftung Rübel	• Berichte des Geobotanischen Instituts, Stiftung Rübel
Ber. Naturwiss.-Med. Ver.	• Berichte des Naturwissenschaftlich-Medizinischen Vereins
Ber. Schweiz. Bot. Ges.	• Berichte der Schweizerischen Botanischen Gesellschaft
Berliner Geogr. Arb.	• Berliner Geographische Arbeiten
Bot. Helv.	• Botanica Helvetica
Bot. Jahrb.	• Botanisches Jahrbuch
Bot. Notiser	• Botaniska Notiser

Bull. Geol.	• Bulletin of Geology
Bull. Pol. Acad. Sci.	• Bulletin of the Polish Academy of Sciences
Butler Univ. Bot. Stud.	• Butler University Botanical Studies
Can. J. For. Res.	• Canadian Journal of Forestry Research
Can. J. Bot.	• Canadian Journal of Botany
Cbl.	• Centralblatt
Conf.	• Conference
Danm. Geol. Unders.	• Danmarks Geologiske Undersøgelse
Dansk Bot. Ark.	• Dansk Botanisk Arkiv
Denkschr. S.N.G.	• Denkschriften der Schweizerischen Naturforschenden Gesellschaft
Diss. Bot.	• Dissertationes Botanicae
Earth Surface Proc. Landf.	• Earth Surface Processes and Landforms
Ecol. Bull.	• Ecological Bulletin
Ecol. Stud.	• Ecological Studies
Eidg. Anst. Forstl. Versuchsw.	• Eidgenössische Anstalt für das Forstliche Versuchswesen
Erdk.	• Erdkunde
Erdwiss. Forsch.	• Erdwissenschaftliche Forschungen
Folia Forest.	• Folia Forestalia
Folia Geobot. Phytotax.	• Folia Geobotanica et Phyto-taxonomica
Folia Quat.	• Folia Quaternaria
For. J.	• Forestry Journal
Forstl. Bundesversuchsanst. Wien	• Forstliche Bundesversuchsanstalt Wien
Forstwiss. Cbl.	• Forstwissenschaftliches Centralblatt
Geobot. Inst., Stiftung Rübel	• Geobotanisches Institut, Stiftung Rübel
Geogr. Ann.	• Geografiska Annaler
Geol. Fören. Stockh. Förh.	• Geologiska Föreningens i Stock-holm Förhandlingar
Geol. Soc. Am.	• Geological Society of America
Geol. Surv. Can.	• Geological Survey of Canada,
Geol. Surv. Finl.	• Geological Survey of Finland
Ges. Vegetationskd.	• Gesellschaft für Vegeationskunde
Ges. Strahlen Umweltforschg.	• Gesellschaft für Strahlen- und Umweltforschung
Global Ecol. Biogeogr.	• Global Ecology and Biogeography

Grana Palyn.	• Grana Palynologica
Int.	• International
Inst.	• Institute
J. Atmosph. Sci.	• Journal of the Atmospheric Sciences
J. Biogeogr.	• Journal of Biogeography
J. Ecol.	• Journal of Ecology
J. For. Res.	• Journal of Forest Research
J. Geophys. Res.	• Journal of Geophysical Research
J. Quat. Sci.	• Journal of Quaternary Science
J. Veget. Sci.	• Journal of Vegetation Science
Jb.	• Jahrbuch
Limnol. Oceanogr.	• Limnology and Oceanography
Medd.	• Meddelse
Medd. Vestl. Forstl. Forsøksstn.	• Meddelelser fra Vestlandets Forstlige Forsøksstasjon
Medd. Nor. SkogforsVes.	• Meddelelser fra det Norske Skogforsoeksvesen
Mitt. Meteorol. Inst.	• Mitteilungen des Meteorologischen Instituts
Mitt. ostalpin-din. Ges. Vegetationskd.	• Mitteilungen der ostalpin-dinarischen Gesellschaft für Vegetationskunde
Mitt. Schweiz. Anst. Forstl. Versuchsw.	• Mitteilungen der Schweizerischen Anstalt für das Forstliche Versuchswesen
Mount. Res. Develop.	• Mountain Research and Development
Münstersche Geogr. Arb.	• Münstersche Geographische Arbeiten
N.G.	• Naturforschende Gesellschaft
New Phytol.	• New Phytologist
N.G.U.	• Norsk Geologiske Undersøgelse
Norges Geol. Unders.	• Norges Geologiska Undersökelse
Norsk Geogr. Tidsskr.	• Norsk Geografisk Tidsskrift
Norsk Geol. Tidsskr.	• Norsk Geologisk Tidsskrift
Norsk Geol. Unders.	• Norsk Geologiske Undersøgelse
Norwegian Arch. Rev	• Norwegian Archeological Review
Philos. Trans. Roy. Soc.	• Philosophical Transactions of the Royal Society
Proc. Bot. Mus. USSR Acad. Sci.	• Proceedings of the Botanical Museum of the USSR Academy of Sciences

Progr. Phys. Geogr.	• Progress in Physical Geography
Quat. Res.	• Quaternary Research
Quat. Sci. Rev.	• Quaternary Science Reviews
Rep.	• Report
Res.	• Research
Rev. Palaeobot. Palyn.	• Review of Palaeobotany and Palynology
S.A.V.	• Slovenské Akademie Věd
Schweiz. N.G.	• Schweizerische Naturforschende Gesellschaft
Schweiz. Z. Forstw.	• Schweizerische Zeitschrift für Forstwesen
Scott. J. Geol.	• Scottish Journal of Geology
Spektr. Wiss.	• Spektrum der Wissenschaft
Suppl.	• Supplement
Sver. Geol. Unders.	• Sveriges Geologiska Undersökning
Ver. Schutze Alpenpfl. Tiere	• Verein zum Schutze der Alpenpflanzen und -tiere
Veröff.	• Veröffentlichungen
Viertelj.schr.	• Vierteljahresschrift
Wiss. Mitt.	• Wissenschaftliche Mitteilungen
Z. Arch. Kunstgesch.	• Zeitschrift für Archäologie und Kunstgeschichte
Z. Geomorph.	• Zeitschrift für Geomorphologie